Computer-Assisted Data Base Design

朱偉恒

一九八二年

Computer-Assisted Data Base Design

George U. Hubbard
IBM Corporation

Van Nostrand Reinhold Data Processing Series

VNR VAN NOSTRAND REINHOLD COMPANY

NEW YORK   CINCINNATI   ATLANTA   DALLAS   SAN FRANCISCO
LONDON   TORONTO   MELBOURNE

Van Nostrand Reinhold Company Regional Offices:
New York   Cincinnati   Atlanta   Dallas   San Francisco

Van Nostrand Reinhold Company International Offices:
London Toronto Melbourne

Library of Congress Catalog Card Number: 80 29519
ISBN: 0-442-23205-5

Manufactured in the United States of America

Published by Van Nostrand Reinhold Company
135 West 50th Street, New York, N. Y. 10020

Published simultaneously in Canada by Van Nostrand Reinhold Ltd.
15 14 13 12 11 10 9 8 7 6 5 4 3 2 1

Library of Congress Cataloging in Publication Data

Hubbard, George U.
  Computer-assisted data base design.

  (Van Nostrand Reinhold data processing series)
  Includes bibliographical references and index.
  1. System design.   2. Data base management.   I. Title.
II. Series.
QA76.9.S88H82   003   80-29519
ISBN 0-442-23205-5

To my Wife and to my Father
for their
Faith, Hope, and Encouragement.

SERIES INTRODUCTION

Data base design has often been done as a black art. The process has been lengthly, obscure, tedious, frought with errors, and sometimes even valueless afterward. Ill-advised choices have been made unwittingly, and their impart not recognized until correction is very costly.

In this book, George U. Hubbard describes ways to improve the process of data base design. A key part of the systematic design procedure he describes is the use of the computer itself to present information helpful to the designer, and reduce the tediousness of the design procedure. The computer also provides helpful reports telling the designer about what data items have been used thus far in the data base design, where some inconsistencies are, and how the design is likely to perform.

In the computer field, we have too often been like the cobbler's children with no shoes. George Hubbard points a way we can in part remedy that situation in the process of data base design.

—NED CHAPIN, PH.D.
Series Editor

THE VAN NOSTRAND REINHOLD DATA PROCESSING SERIES

Edited by Ned Chapin, Ph.D.

Preface

This book presents a methodology for data base design in which systematic techniques are organized into a series of iterative procedures with numerous decision points. Beginning with the gathering of the initial data requirements, the process is carried through the design and evaluation of the logical and physical models. In addition to presenting a design methodology, the book focuses on the possible automation of major portions of that methodology.

At the current state of the art, one giant design program is not being advocated because much human interaction is still required. The trend toward computer-assisted data base design has been gradual and piecemeal because of the complexity of the process. What is needed now is an overall design methodology into which current and future computer-assisted techniques can fit in a consistent and harmonious manner. This book provides such a methodology and suggests many of the techniques that can be used. Some of these techniques are based on concepts that are embodied in two IBM program products, Data Base Design Aid and DBPROTOTYPE II (References C.3 and D.4); however, this book is by no means limited to the functions or scope of those programs, nor is it intended as a description of them. While the emphasis is on automating as much of the design process as possible, the material in this book is presented in such a way as to be instructive and useful regardless of whether or not computer assistance is used.

In current practice, data bases are normally designed by data processing personnel specializing as Designers, and they are designed on the basis of data requirements originated by End Users representing the applications to be processed. The design concepts to be presented can help end users understand better how to specify their data requirements. By following procedures based on these concepts, designers can perform a more thorough and consistent analysis of the data requirements and their inevitable design trade-offs. Communication between designers and end users can become more meaningful and complete, and design evaluation and control by management can be more objective. An additional by-product of these procedures is that designers and end users can obtain a more objective basis for dealing with the differing interdepartmental perspectives that seem always to beset a design study.

Therefore, an understanding of the concepts and procedures of a systematic approach to data base design can benefit more than one audience. This book should be of value to the manager who wants to increase his understanding of the concepts of data base design, to the practitioner who wants to improve his design skills (with or without computer assistance), and to the end user who initially specifies the data requirements and who needs to communicate fluently with the data processing personnel.

This book was written while the author was on a sabbatical assignment at IBM's Systems Research Institute in New York City. The author is currently with IBM in Dallas, Texas. The views expressed in this book are those of the author and do not necessarily represent the views of IBM. The author assumes full and sole responsibility for its contents.

The author is indebted to the following people for reviewing the material and for making many helpful suggestions: Mark Gillenson, Okan Gurel, Darrell Jones, and Alan Parnass of IBM; Jeff Oakes, Pershing Parker, and Tom Ruotolo of the Hartford Insurance Group.

GEORGE U. HUBBARD

Contents

PART III LOGICAL DESIGN

PART V Ancillary Design Considerations

Computer-Assisted
Data Base Design

PART I
INTRODUCTION

1. Overview

THE DATA BASE DESIGN PROBLEM

Data base design can be a long and tedious process. Designing large data bases, say of 500 elements or more, usually takes months. In some cases, years have been expended. Until now, data base design has been more an art than a science, with no standardly recognized set of rules or procedures. Experience and intuitive feeling have been the designer's main resources.

In addition to the time and tedium involved, the quality of the resulting design is frequently a problem. In many poorly performing systems, the data base design has been found to be the major culprit. Determining the real meanings of the data elements is one of the major design problems. But once this is done, well-designed structures can still be difficult to obtain because of the size and complexity of the structuring process. Function and performance are always major design criteria. The data base must be able to provide the necessary data with reasonable processing efficiency. Further, a poor data base design can make application programming more difficult by requiring more complicated programming logic than is necessary for accessing the desired information. This could result in increased time and expense for implementation and increased processing time after implementation.

Integrity and consistency are also vitally important. Data must not be accidently lost or destroyed, and replications of the same data must have consistent values within controlled limits. What the user receives should be consistent with his expectations. Privacy and security are other important design criteria. The data must be protected against unauthorized access. Extensibility is another item of major importance. The data base must be able to grow and to change with changing requirements without unduly disrupting existing applications. These and other design considerations usually pose conflicting constraints, and a systematic design procedure can help the designer in recognizing and resolving many of these conflicts before the data base is installed.

One of the more serious consequences of poor data base design is the frequently prohibitive cost of restructuring an existing data base once it has been determined to be deficient. With regard to DL/I data bases, for example, changes to the logical design (i.e., hierarchical structure, segment size and content, logical relationships, and secondary indexes) usually mean changes to the application programs as well as reorganization of the data base itself. The motto here is "get it right the first time." Physical design changes (access method, block size, secondary groups, pointer options, etc.) are usually less severe in their impact. In these cases, applications may not have to be changed, although reorganization is usually required. A few design parameters are relatively innocuous. Such things as insert-replace-delete rules, data base size, physical placement of indexes, etc., normally do not require program changes or reorganization. On the other hand, application program logic could be sensitive to any of these criteria, and could have to be changed to support corresponding data base changes. A systematic approach with thorough analysis techniques can help reduce the likelihood of such changes.

Traditionally, the roadblocks to good data base design fall into the following categories:

- Insufficient analysis of data requirements (including naming semantics and data relationships)
- Lengthy structuring tedium beyond human capacity for manual analysis
- Difficulty of predicting performance implications
- Management-imposed time constraints

Systematic data base design procedures augmented by computer-assisted techniques can make significant contributions toward alleviating these problems.

THE DESIGN GOALS

A data base design methodology of rules, techniques, and procedures can be formulated which, if properly applied, can help shorten the design cycle and improve the design quality. Using automated techniques, design information can be made available earlier in the design cycle and in a more complete manner than is usually obtained manually. With manual techniques, it is often difficult for the designer to "see the forest for the trees," or even for him to tend to all the "trees." Computer-assisted design techniques can help fill this gap.

The goals, then, of computer-assisted data base design are to:

- Improve the design quality
- Shorten the design cycle

More specifically, the goals are to produce a well designed and well tuned data base with appropriate extensibility and integrity features and to do it in a minimum amount of time.

THE PROPOSED METHOD

Significant portions of a systematic design procedure can be automated to provide helpful design information while avoiding much of the tedium normally associated with data base design. A variety of Editing Reports can be provided to assist the designer in standardizing the names and meanings of the data elements. Diagnostic reports from the logical structuring process can reveal inconsistencies, omissions, and alternatives. Performance estimates and measurements help the designer evaluate various possible design variations. And up-to-date documentation can be made available throughout the entire process.

The data base design procedures to be presented are shown in Figure 1-1. The process is organized into three design phases. Conceptual Design means gathering, recording, and editing the data requirements. Logical Design organizes the data requirements into structures (e.g., segments and hierarchies) for a given data base management system. Physical Design deals with storage patterns, access methods, and performance-oriented options.

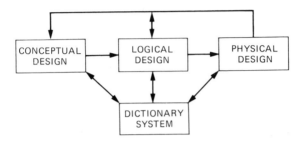

Figure 1-1. Data base design phases.

Each design phase is considered to be an iterative set of procedures, the results of which will be a *model*. Thus, we will speak of converging onto a conceptual model in the conceptual design phase, converging onto a logical model in the logical design phase, and converging onto a physical model in the physical design phase. In addition, the overall process is *iterative*. Results may be obtained in the logical design phase or in the physical design phase that will require changes or reinterpretations in earlier phases to obtain required capabilities.

Figure 1-1 also indicates desirable interfaces between each of the design phases and a Dictionary System. The nature and benefit of these dictionary interfaces are suggested in Chapter 21. However, the design procedures to be presented can be used independently of a dictionary system, and are applicable to designers in nondictionary environments.

While many steps of the design process lend themselves to automation, considerable human interaction is required. The iterative procedures require frequent dialogues between the designer and the end users, and well-designed automation can provide insights and suggestions of specific problems to be discussed and resolved in these dialogues. Figure 1-2 (from which the possible

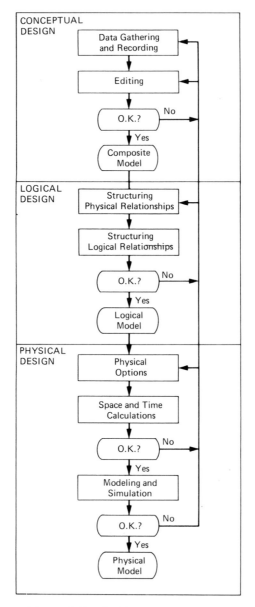

Figure 1-2. Detail of data base design phases.

dictionary interfaces are omitted) depicts the iterative procedures in slightly more detail along with the major human decision points.

THE EXPECTED BENEFITS

The data base design approach to be presented can help the designer achieve the goals listed above by providing a methodology and a set of design techniques. Many of the techniques can be implemented by computer programs to relieve the designer of much of the drudgery of the design process and also contribute to a more thoroughly analyzed design. When used as intended, the methodology and the techniques can provide the following benefits:

- Improve the design quality
  - —Force a more thorough analysis of the application specifications
  - —Reveal errors, inconsistencies, and omissions in the data requirements
  - —Remove nonessential (transitive) associations
  - —Identify redundancies
  - —Identify structural design alternatives
  - —Analyze performance consequences
  - —Provide convenient extensibility studies
- Shorten the design cycle
  - —Reduce the number of design iterations
  - —Reduce the time for each iteration
  - —Document the results

SCOPE AND ORGANIZATION OF THIS BOOK

Although the data base deign procedures to be presented can apply, up to a point, to any of the three data base organizations (hierarchical, network, and relational), the major emphasis in this book will be toward the hierarchical structures embodied in IBM's Information Management System (IMS) (References A.4, A.5, A.10, D.1, and D.5). Conceptual design is independent of any data base structure, and the techniques to be presented can apply to any data base management system. While much of the logical design process is also generally applicable, the resulting model will be structural. The major emphasis in this area will be designing for IMS-type hierarchical structures, although separate chapters will indicate the extensions of these procedures to the relational model and to the network model as specified by CODASYL (References A.4, A.5, A.8, A.9, and A.10). The treatment of physical design will be very specifically oriented to IMS structures.

The starting point for this book assumes that the enterprise has already been analyzed and that the functional requirements for the applications to use the data base have already been determined. The task at hand is to derive and record the data requirements from those applications, organize them into an efficient data base design, and document the progress and results. Thus, conceptual design will be treated in a somewhat limited sense. It will be oriented primarily toward philosophies of data requirements gathering and toward editing the requirements for consistency in data usage and in naming standards. Logical design and physical design will be treated in full. Finally, suggestions will be offered regarding desirable interfaces between the design process and a dictionary system, and also regarding an interactive data base design language.

To illustrate the procedures of computer-assisted data base design, a case study will be included as an appendix. This case study will illustrate the design of a data base for a football league. Throughout the text, numerous examples from this case study will be used to illustrate the concepts being explained. When studying these individual examples, the reader may occasionally want to refer to the case study itself to see them in the context of the complete set of application data requirements.

Finally, it should be pointed out that while the procedures to be presented are one approach to data base design, they are not the only way. For example, instead of specifying functional data requirements in terms of binary relations and association types (as we shall do) between individual data elements, another approach is to determine the entities (objects, concepts, etc.) that are involved and about which descriptive data is to be stored, merge these object-level entities and their apparent relationships into a structure, and then modify that structure as necessary to accomodate the functional requirements (References C.12 and C.13). Although the procedures to be presented are described in terms of the former approach, they can be directly applied to this latter approach as well. There are other ways to derive the data relationships. One approach, item level synthesis (References C.9, C.10, and C.11), statistically analyzes which data elements are used together most often by the various application functions and thus derives the logical data structures. While there are different starting points for data base design, the one to be described is valid and has been proven by experience.

2. Basic Definitions

Because of the great diversity of terms and their various meanings in the data base area, some basic definitions are in order at this point. Further definitions will be presented, as needed, as the design concepts and procedures are addressed. An attempt will be made to adhere as closely as possible to DL/I and to ANSI-SPARC terminology.

DL/I (Data Language/One) is the name generally used to refer to the data structures of IMS (References A.4, A.10, D.1, and D.5). More precisely, however, it refers to the interface language used by application programs to invoke the IMS services. We will use DL/I in its broader sense of referring to the data structures themselves. These structures, which appear as a hierarchical tree to the application program, are actually implemented as restricted networks.

ANSI-SPARC refers to the work of the Standards Planning and Requirements Committee of the American National Standards Committee on Computers and Information Processing which was established to see which, if any, of the various data base management systems are suitable candidates for standardization (Reference A.2).

ELEMENTARY TERMS

Enterprise

An Enterprise is that part of the real world being described or modeled by the data base and its applications. An enterprise may be an entire company, a functional area within a company, a process, a system, and so forth. It is modeled in terms of its entities (objects or concepts) and the functions performed by or on these entities.

Entity

An Entity is an identifiable object, concept, or activity belonging to the enterprise and about which descriptive data is recorded. Entities can be people, parts, accounts, skills, and so forth. An entity is described in terms of selected properties to which values (words, numbers, or codes) are assigned.

Data Element

A Data Element is a property of an entity and is represented by a name and by a set of values. Data element names can be EMPLOYEE-NUMBER, BIRTHDATE, PRICE, BANK-BALANCE, SKILL-LEVEL, and so forth. The values assigned to a data element describe the represented property for each entity occurrence being considered or are used to identify the entity occurrence. The data element names are usually stored as part of the data base description (schema), while the values are stored in the data base itself. Data elements are frequently referred to as fields which serve as keys or as attributes.

Entity Record

An Entity Record is the set of data element values that describes a particular entity occurrence. Assuming that a PART entity is to be described by data elements named PART-NUMBER, PART-NAME, MATERIAL, and COST, an entity record for a particular part might be AX1234 WASHER BRASS $1.98. An entity record is analogous to a segment occurrence in DL/I, to a tuple in relational structures, or to a record occurrence in CODASYL.

Entity Record Set

An Entity Record Set is the set of entity records for all occurrences of the entity type being considered. An entity record set is analogous to the set of occurrences of a segment type for a given parent occurrence in DL/I. It is also analogous to a relation in relational structures and to the members of a CODASYL set.

Identifier

An Identifier is a data element (or a combination of data elements) whose value(s) may be used to determine one or more values of a related data element. The identification can be Unique (single-valued) or Nonunique (multi-valued). If unique, the identified element (attribute) is said to be functionally dependent on the identifying element (key).

Key

A Key is an identifier that uniquely identifies an entity record. As such, it uniquely identifies the attribute values for that entity record. In an entity record set, it is possible to have more than one data element whose values uniquely identify the entity record. Each of these data elements is a Candidate Key, and one is usually chosen as the Primary Key. A primary key may be implemented as a unique sequence field in a DL/I segment.

Attribute

Data elements that are not primary keys are referred to as Attributes. In an entity record, the attribute values are identified by the primary key value. An attribute is analogous to a nonsequence field in a DL/I segment.

RELATIONSHIPS

Mappings

Mappings have been the traditional vehicles for indicating the nature of the relationships between pairs of related data elements. A brief review of mappings is given here.

One-to-One Mapping

A given occurrence of the "from" element identifies one and only one occurrence of the "to" element, and vice versa. The identification is unique in both directions. The example in Figure 2-1 is EMPLOYEE-NO and SOCIAL-SECURITY-NUMBER. Each uniquely identifies the other.

EMP-NO   SOC-SEC-NO

Figure 2-1. 1:1 mapping.

One-to-Many Mapping

A given occurrence of the "from" element identifies any number (zero, one, or more than one) of occurrences of the "to" element. The identification is not necessarily unique in that direction. But in the inverse direction, a given occurrence of the former "to" element identifies one and only one occurrence of the former "from" element. The identification in this inverse direction is unique. The example in Figure 2-2 is DEPARTMENT-NO and EMPLOYEE-NO. A given department normally has many employees, but each employee belongs to only one department.

Figure 2-2. 1:M mapping.

Many-to-One Mapping

This is the same as one-to-many mapping and is illustrated in Figure 2-3. The relationship is associative.

Figure 2-3. M:1 mapping.

Many-to-Many Mapping

A given occurrence of the "from" element identifies any number of occurrences of the "to" element, and vice versa. The identification is nonunique in both directions. The example in Figure 2-4 is PART-NO and SUPPLIER. A given part may be furnished by many suppliers, and a given supplier can furnish many parts.

Figure 2-4. M:M mapping.

Associations

Mappings describe two-way relationships between pairs of related elements. It is also important to deal with Associations, which are the one-way relationships between pairs of related elements. Between a pair of related elements, the associations in both directions (the forward and the inverse), if both are defined, constitute a mapping.

It is frequently desirable to express data relationships in terms of one-way associations, as well as by complete mappings, and structuring processes should be able to deal with this practice. To one user, the nature of some of the inverse associations may not be important and arbitrary choices may conflict with the more firm requirements of other users. Also, associations (as will be seen) are used to define key-to-attribute relationships and, in most cases, relationships are not specified from attributes back to their keys. Therefore, the design procedures to be presented are based on data relationships specified in terms of association types rather than in terms of mappings.

Three types of associations are defined as follows:

- Type 1 (simple)
- Type M (complex)
- Type C (conditional)

Type 1 (Simple) Association

A given occurrence of the "from" element identifies one and only one occurrence of the "to" element. The identification is unique (single-valued and atomic), and it represents a functional dependency. One example in Figure 2-5 is EMPLOYEE-NO to SOCIAL-SECURITY-NUMBER. Another example is EMPLOYEE-NO to DEPT-NO. An employee has only one social security number and belongs to only one department. Nothing is said about the nature of the relationship in the inverse direction.

Figure 2-5. Type 1 (simple) association.

Type M (Complex) Association

A given occurrence of the "from" element identifies any number (zero, one, or more than one) of occurrences of the "to" element. The identification is not necessarily unique, and it represents a multivalued determination. In Figure 2-6, an example is DEPT-NO to EMPLOYEE-NO. Another example is

Figure 2-6. Type M (complex) association.

PART-NO to SUPPLIER. A given department may have many employees, and a given part may be furnished by many suppliers. Again, nothing is said about the nature of the inverse association.

Type C (Conditional) Association

For a given occurrence of the "from" element, a corresponding occurrence of the "to" element may or may not exist, but if it does exist, there is only one. The identification, if made, is unique. An example of EMPLOYEE-NO to TERMINATION-DATE is shown in Figure 2-7. Another example is HOS-

Figure 2-7. Type C (conditional) association.

PITAL-BED to PATIENT. An employee may or may not have a termination date, but if he has one, he has only one. A hospital bed may or may not have a patient assigned to it, but if there is a patient then, hopefully, there is only one.

PART II
CONCEPTUAL DESIGN

Conceptual design is the task of determining an enterprise's information requirements and the processes and data required to provide that information. In its broadest sense, conceptual design begins with a study of the nature of the business of the enterprise. Its goals and objectives are defined, and the processes (operational, control, and planning) for achieving those goals and objectives are analyzed. It then identifies successive sublevels of these processes (administration, marketing, payroll, shipping, etc.) until each process is decomposed into applications and functions to be implemented without further subdivision. The information requirements of these applications are also derived and data requirements are determined for those applications to be automated.

For the purposes of this book, conceptual design will be considered in a much narrower context. This treatment of computer-assisted data base design will assume that the application functions to be automated have already been determined. This in itself is an exercise for which computer-assisted techniques may be quite helpful, and is the subject of considerable current research. The remaining portion of conceptual design is to determine the data requirements (the data elements and their relationships) of these functions, and to provide consistency in the interpretation and intended use of these requirements. In order to do this properly, the designer must have an understanding of the nature of the business and the procedures used by the enterprise being modeled. These considerations will be addressed in the next three chapters.

Chapter 3 will discuss the various perspectives from which data can be viewed and will define the Composite Model which is the subset of the conceptual model that represents the data to be actually stored in the data base. The Conceptual Model, on the other hand, is a graphic representation of all of the data requirements, stored and nonstored, automated and nonautomated. Methods of recording data requirements are presented in Chapter 4. Various sources for gathering the requirements are considered along with considerations for interpreting and recording special situations. In Chapter 5, data editing is discussed and possible editing reports (their content and usefulness) are outlined.

3. Views of Data

WAYS OF VIEWING DATA

It is generally agreed in the data base community that there are three realms in which an enterprise, or the real world in general, can be viewed (Figure 3-1). There is the enterprise (1) as it actually is, (2) as it is perceived by humans, and (3) as it is described by symbols. In this context, we can say that we are dealing (1) with reality, (2) with a descriptive representation of that reality, and (3) with data that characterizes that representation. It is this characterizing data, or portions thereof, that is stored in the data base and manipulated by the application programs.

With regard to application programs and data storage, it is also generally agreed that the data used to characterize an enterprise can be viewed from three perspectives or views (Figure 3-2). These views are most commonly called the Conceptual View, the External View, and the Internal View.

External View

The external view of the data represents the data requirements of a given function (or program). Thus, there is an external view for each function. (Later in this chapter, this definition will be slightly expanded.) The various external views usually overlap to some extent but are rarely identical.

It will be convenient to divide the concept of external view into a Perceived External View and a Local External View. Starting with the perceived external views, the end users and the data base designers will derive the local external views as input into the design procedures.

Perceived External View

For a given function, the perceived external view (hereafter called the Perceived View) is the set of data requirements visibly implied by functional specifications or, more directly, by actual output and input formats. The perceived view of a function represents what the users actually want to receive from the function and also what they expect to supply to it.

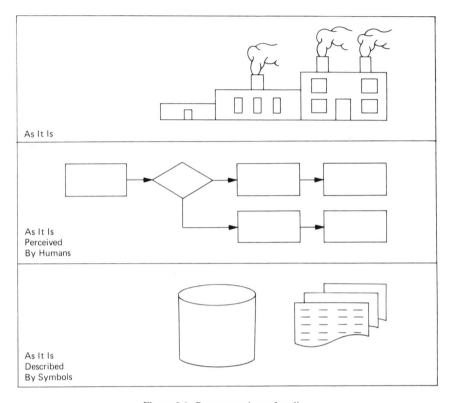

As It Is

As It Is
Perceived
By Humans

As It Is
Described
By Symbols

Figure 3-1. Representations of reality.

Local External View

The local external view of a given function (hereafter called the Local View) represents the data elements and relationships actually required in the integrated data base to support the function. The local view roughly corresponds to the Program Specification Block (PSB) in DL/I systems.

There is not always a one-to-one correspondence between a perceived view and its local view. For example, the perceived view may contain data (NET-PAY, AGE, CURRENT-DATE, CURRENT-SALARY, etc.) not desired for storage in the data base but which can be materialized from other data that is in the data base. On the other hand, data not noted in the perceived view, but directly related to it (GROSS-PAY instead of NET-PAY, BIRTHDATE instead of AGE, etc.), may be required in the data base to support the proposed function. Additional data, not directly related to that of the perceived view (tax tables, Julian calendar, statistical distributions, data provided by other functions, etc.), may also be required in the data base to accomplish the processing

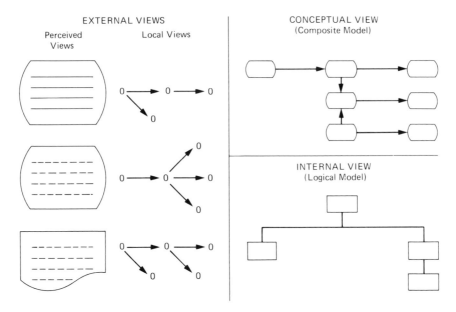

Figure 3-2. Views of data.

functions. Bridging the gap between the perceived and local views is a very important part of the designer's task of gathering and recording the data requirements.

Conceptual View

The conceptual view of data represents the entire set of data requirements as derived from the perceived views of reality. In its broadest sense, the conceptual view can be construed to represent the total information requirements of the enterprise being modeled. This can include nonstored data for manual procedures as well as data to be included in the data base. For our purposes, the conceptual view will be considered in the more limited sense of representing the total collection of data elements and their relationships to be stored in the data base being designed or to be materialized by the application programs. In this sense it represents the composite of the external views for which the data base is being designed. There is only one conceptual view for the enterprise being modeled and it is nonstructural with regard to networks, hierarchies, or relations.

The conceptual view serves as the basis from which the integrated data base is designed and from which the perceived views are materialized. One of its purposes, therefore, is to assure an integrated data base capable of accepting and providing the desired external views via the application programs. Consisting (in our limited sense) of the composite of the external views, the conceptual view is represented by a composite model which is derived by merging the several local views and which is operated upon to obtain the internal view. The conceptual view also serves as a standard against which additional external views can be evaluated to determine if they are consistent with the data requirements and standards already established.

Internal View

The internal view of the data represents the integrated data base itself. The internal view is roughly analogous to the Data Base Description (DBD) in DL/I systems. In logical design it is represented by a logical model showing the data organized into the structures of a given data base management system. In physical design, a physical model shows how these structures are implemented onto the physical storage devices.

In content, the internal view is generally a subset of the conceptual view. Rather than being stored in the data base, some of the data in the conceptual view may be materialized at execution time as the result of calculation. Some may be relegated to manual processing. Or some may remain (or be placed) on tape separate from the data base. The intent of the internal view is to define all the data (and only that data) required by a given data base management system to support the functions that will use the data base.

The task of obtaining a suitable internal view (integrated data base) is one of the primary subjects of this book. A practical internal view is not merely a subset of the logical sums of the initial local views. When local views come from different departments having different habits and jargon, conflicts in their respective local views usually exist and must be detected and resolved. Conflicts within local views frequently exist also. The procedures to be presented are intended, in large part, to assist the designer in finding and resolving these conflicts.

SOURCES OF LOCAL VIEWS

The task of the designer in the conceptual design phase is to specify local views for all the functions that will use the data base being designed. One of the major purposes of the automated procedures to be described is to help the designer specify these views correctly and consistently.

Two primary sources for obtaining local views are suggested: the Functional Requirements and the Intrinsic Requirements.

Functional Requirements

The data requirements of each function that will use the data base are specified. The data elements of each function are identified with the association types between pairs of related elements.

From Perceived Views of New Functions

Each new function (or old function being redesigned for a data base environment) will exist to produce some kind of result, and will require some kind of input to drive it. What output units (reports, screens, tables, records, etc.) are produced? What input units (messages, cards, files, etc.) are needed to drive them? What additional data (tables, existing files, computational results, etc.) is needed by the processing? For each of these output, input, and processing units, the data elements involved are identified and the relationships between them are specified. Starting with the perceived views represented by actual output and input layouts or by functional specifications, the question is: What data are needed in the actual data base to support the perceived views? This is the process of deriving the local views.

From Existing Programs

In some cases existing programs are used in a data base system without redesigning their functional logic. Either they work sufficiently well and management is satisfied or time constraints do not permit the redesign. A frequent problem is that noone knows their exact data requirements. Either the original programmers have left or documentation is missing or both. An automated scanning program can determine the data requirements for a program with an intact source code. By examining the declarations, the formats of the input/output (I/O) work areas, the move instructions, and the results of calculations, the required data elements and their relationships can be derived. The danger here is in preserving an application design that is probably not well suited for a data base environment, and this source of local views, if used at all, should be used with caution.

Intrinsic Data

In addition to supporting the functional requirements, data are often considered resources representing organization and operation of an enterprise and, as such, the data have meaning and relationships regardless of the functions using them. For example, in an electric power company a major entity type is wire.

The wire is characterized by its material, diameter, weight, insulation, cost, etc. It is furnished by certain suppliers, stored in certain quantities at certain locations, used in certain configurations, etc. Intrinsic data and relationships exist simply because of the nature of the business and the physical relationships and they generally exist quite independently of any application program requirements. To a large extent, these intrinsic relationships form a framework within which the functional data requirements must fit. Therefore, the definition of external views given above should be slightly enlarged to include the concept of intrinsic data requirements as well as individual functional data requirements.

From Perceived Views

The perceived views of the intrinsic data are usually found in such documents as blueprints, engineering specifications, legal regulations and constraints, corporate policies, etc. In practice, they may also be obtained simply from the minds of some of the more experienced personnel. Again, the fundamental question is: What data is needed in the actual data base (local views) to support these perceived views? With intrinsic data, the local views and the perceived views are more likely to be identical than with functional requirements.

From Existing Files

Sometimes, time constraints dictate the direct migration of existing files (particularly ISAM data sets) to the data base system without restructuring. The fact that the data and its relationships already exist and are to be used "as is" causes such files to be considered also as intrinsic data. For data requirements obtained in this manner, the data base management system is being used as little more than just another access method. Good Indexed Sequential Access Method (ISAM) structures may not necessarily correspond to good data base structures. In addition, poor designs may be propagated into the data base by this approach. Again, caution and restraint should be exercised when considering this type of source.

APPROACHES FOR SPECIFYING LOCAL VIEWS

There are differences of opinion over whether the functional requirements or the intrinsic requirements should play a dominant role in requirements gathering. The purpose of the data base is to support the functions to be processed, and unless it is designed according to the functional data requirements, it may not provide the required support. On the other hand, the intrinsic data describes the enterprise as it is perceived to exist and, unless it serves as the design framework, the functional requirements may prove to be unrealistic and unsupportable.

Computer-assisted techniques support an iterative buildup of the data requirements and make the following approach possible. (The iterative nature of these automated procedures is more fully discussed in Chapter 12.) Beginning with functional requirements (for the sake of example), the following iterative approach is suggested for combining local views and for converging onto a desirable logical design.

Select the one or two most important applications or functions, derive their local views, put them into appropriate machine–readable formats, and invoke the conceptual and logical design procedures to derive a logical model. The result is a logical design that directly supports the "more important" functions. Now select the next most important applications (or functions), add their local views to those already obtained, and rederive the logical model. If incompatabilities exist between the old and new logical models or if new diagnostics occur, these differences and diagnostics are the result of adding the new requirements and required trade-offs can be identified and evaluated intelligently and in a controlled manner between the "greater" and "lesser" requirements. The intrinsic requirements, as far as they can be determined, should be included at some appropriate point in this iterative procedure to assure that the functional requirements are compatible with the intrinsic. Any necessary trade-offs between functional and intrinsic requirements can be readily recognized and evaluated in this manner. To complete the process both functional and intrinsic future requirements should be estimated and processed to maximize the likelihood of obtaining an extensible data base structure. This iterative approach, which is usually not feasible by using purely manual design techniques, can provide a significant quality control capability to the design process.

4. Gathering and Recording Local Views

The objective of gathering the data requirements is to obtain a set of local views that can be combined (by the procedures to be described) into a composite model for editing (from which a logical model can be derived). Assuming it has already been decided which portions of the enterprise are to be served and which application functions are to be implemented, we begin by examining the functional specifications and their data requirements. From these requirements, local views are to be derived.

DETERMINING FUNCTIONAL REQUIREMENTS

Within the scope of this book, automated design procedures can make their initial contribution by helping the end user develop his functional specifications. Although functional specifications are often "fully" defined before entering the design process, it is also true that they frequently are changed during the course of the design process as the users realize that what they initially specified (or thought they wanted) no longer corresponds exactly with what they now realize they need. By producing exact formats of the desired screens and reports, computer assistance can help the end user more clearly see how well the expected reports will meet his real information needs. And an online query capability into a sample data base can further assist the end user in defining his processing requirements. With these types of assistance, the end user knows that what he now sees is what he will eventually be getting from the system. Experience indicates that when this approach is used the data requirements are less likely to change during the design process.

As an example of finalizing functional specifications and determining data requirements, the following sample dialogue illustrates how the data base designer and the end user, or applications specialist, can work together using this type of computer assistance. Assume that a data base is being designed for a football league, and the designer is interviewing the player personnel officer (PPO) of one of the teams to help him specify his requirements.

PPO: I need a report showing what players are playing which positions, and for each player I want to see his size, experience, and ranking.

Designer: You mean a player can play at more than one position?

PPO: That's right. If someone gets hurt we have to fill in from somewhere.

Designer: And by "size" I presume you mean weight and height.

PPO: That's right.

Designer: O.K., I think I know what you want. Give me a few moments, and I'll show you what it might look like.

The designer keys in some keywords and descriptors at his terminal and receives back the sample report in Figure 4-1.

Designer: How does this look to you? I included "age" thinking that might also be of interest.

PPO: Good. I need "age" but the report is hard to read. It's not obvious who our best players are.

Designer: I spent yesterday with the coaches, and this is the type of presentation they like.

PPO: I know, but their interests are different from mine. They are interested primarily in position, what players are available for each position, and how they are evaluated by position. But I have to pay these players and I need an overall team evaluation. And put in "player number" also. That helps me know who they are.

	BISMARCK BENGALS PLAYER—POSITION LIST					
POSITION	PLAYER	AGE	WEIGHT	HEIGHT	EXP	RANK
Split End	Barry	25	210	6– 3	1	2
	Jones	27	185	6– 1	3	1
	Riley	32	190	6– 2	5	3
Tight End	Duey	26	195	6– 4	4	1
	Hall	23	215	6– 4	1	3
	Johnson	21	200	6– 3	2	2
Flanker	Eccles	35	190	6– 1	3	4
	Hall	23	215	6– 4	1	3
	Riley	32	190	6– 2	5	1
	Welter	30	180	5–11	1	2

Figure 4-1. Sample report (1).

Designer: I suspect that when you use the term "rank" you mean something different from the "rank" the coaches use.

PPO: That's right. They are talking about rank by position, and I need to see an overall team ranking.

Designer: Now we're getting somewhere. But the coach uses "jersey number" and you have mentioned "player number." Are you both talking about the same thing?

PPO: Yes. We need to get together on our names.

Designer: I'll make a few alterations. Now, let's see if this revised report suits you better.

PPO: That's more like it. Now I can see the information I need. Now, let's add. . . .

BISMARCK BENGALS
TEAM PLAYER RANKINGS

TEAM-RANK	P-NO	PLAYER	POSITION	POS-RANK	AGE	WEIGHT	HEIGHT	EXP
1	85	Jones	Split End	1	27	185	6– 1	3
2	81	Riley	Flanker	1	32	190	6– 2	5
			Split End	3				
3	47	Welter	Flanker	2	30	180	5–11	1
4	80	Duey	Tight End	1	26	195	6– 4	4
5	86	Johnson	Tight End	2	21	200	6– 3	2
6	83	Barry	Split End	2	25	210	6– 3	1
7	49	Hall	Tight End	3	23	215	6– 4	1
			Flanker	3				
9	45	Eccles	Flanker	4	35	190	6– 1	3

Figure 4-2. Sample report (2).

As sketchy as this dialogue is, it is indicative of the interaction that must take place between the data base designer and the end user at some point in the design process. When this interaction occurs early so that the end user can see and study the "actual" results of reports and sample queries, the specification of functional requirements can usually be completed much sooner and with less likelihood of revision after structuring begins.

DERIVING AND RECORDING LOCAL VIEWS

Describing Local Views

Local views are described in terms of Binary Relations (i.e., pairs of data elements and the one-way association types relating them) which may be recorded as graphs, matrices, or interconnected lists. For example, in a football league, suppose a team roster is wanted in which each player is identified by a player number, and the information to be recorded about each player is name, age, height, weight, and position. Once a given player is identified, his name, age, height, and weight are uniquely identified. Therefore, there is a binary relationship with a Type 1 association between player number, an identifier, and each of these four uniquely identified attributes. Graphically, these binary relations can be depicted as follows. In this book, we will refer to these graphs of binary relations as *bubble charts*. The bubble chart of the relations just discussed is shown in Figure 4-3.

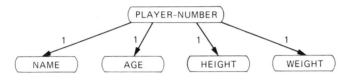

Figure 4-3. Bubble chart.

Depending upon the intentions of the end users, there may be different relationships between a player and his position. If the user means the one position the player currently plays, then the identification is unique and it is also represented by a Type 1 association. But if the user means all the positions at which the player might be used, the identification is nonunique and is represented by a Type M association. Let's assume for now that most players can play several positions, and we want to identify all of them. All the binary relations are now depicted graphically as the bubble chart of Figure 4-4.

This represents the perceived view of the data required for the Player Roster.

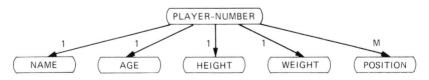

Figure 4-4. Perceived view.

But the designer, or end user, should really be trying to specify the local view, which, as previously defined, may not be identical to the perceived view. The difference between these two views is important, though sometimes subtle. The designer should be asking: What data is needed in the data base to support the proposed function? Age is generally not a data element to be stored in the data base because it is subject to frequent changes, especially if there are a lot of players. Birthdate, from which age can be calculated, is a much more stable element. Therefore, the designers elect to use birthdate rather than age, and the perceived view of Figure 4-4 becomes the local view illustrated in Figure 4-5.

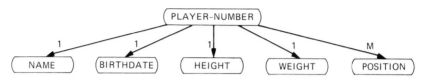

Figure 4-5. Local view (1).

Now, let's assume that the coach also requires a record of each player's experience and skill at each of the positions at which he can play. The binary relations for this new local view may be those shown in Figure 4-6.

The individual local views are submitted to the automated procedures for initial processing which consists of editing and combining them into a composite model for logical structuring.

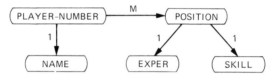

Figure 4-6. Local view (2).

The Composite Model

The data base designer, or end user, provides the various local views as primary input to the automated procedures which will combine the local views into a composite model. Figure 4-7 shows the composite of the two local views of Figures 4-5 and 4-6.

A preliminary editing operation is performed on the individual local views to assure they are specified consistently and completely. Then after constructing the composite model, full-scale editing is performed. The edited composite

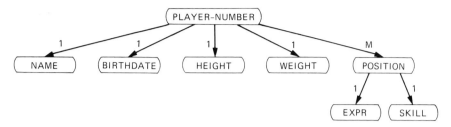

Figure 4-7. Composite model.

model represents the unstructured internal view of the integrated data base; from it a logical model, and then a physical model, will be derived according to the rules of the data base management system to be used.

Deriving Binary Relations

Deriving the appropriate binary relations for an intended function will sometimes pose difficulties of interpretation. Consider a requirement in which, for a given employee, reference is to be made to the city, county, and state in which he lives or works. The designer might be inclined to specify the relationships linearly as shown in Figure 4-8.

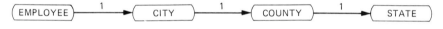

Figure 4-8. Linear Type 1 associations.

The rationale might be that an employee lives in one city, that city is in one county, and that county is in one state. It will be shown in Chapter 9 that data elements related in this manner might be structured into separate segments on a hierarchical path, and this may not be what the designer wants.

A different viewpoint is expressed with radiating Type 1 associations as shown in Figure 4-9.

In this case the thinking is that EMPLOYEE is the sole identifier, and that

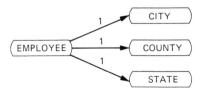

Figure 4-9. Radiating Type 1 associations.

for a given employee, there is a city, a county, and a state. There is no attempt here to define data about a city, county, or state. These data elements are considered to be merely attributes of the employee entity. Unless requirements from other local views dictate otherwise, this view will produce an employee segment with city, county, and state as fields in that segment.

If, on the other hand, it is desired to store information about the cities, counties, and states, as separate entities, a variation of the view of the bubble chart in Figure 4-8 may be most appropriate.

Figure 4-10 represents a locality file in which data is required about the states, about the counties in each state, and about the cities in each county. Here these three data elements represent entities in their own right rather than being attributes of the employee entity. Employees can be related to these entities according to the requirements of the application functions.

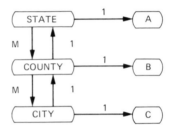

Figure 4-10. Specifying entities.

Determining Association Types

The correct specification of association type is not always obvious, and it requires a thorough understanding of the intent of the application program's use of the data and sometimes an understanding of the resulting structural implications. Although we would like to avoid thinking of structural implications when specifying the data requirements, it is not always possible or practical to do so.

Starting with a simple example, consider the previously mentioned association from PLAYER to POSITION. The association is Type 1 if thinking of a unique position, such as current position, for a given player. If thinking of the various positions a player can play, or has played, the association is Type M. For an additional perspective, consider the data elements CITY and STATE. Thinking of a given state and dealing with the characteristics of its cities is very different from thinking of a city, such as Springfield, and dealing with all the states containing a city by that name (Figure 4-11). In the latter case, there is a Type M association from CITY to STATE.

In the former case, that of the cities in a given state, a Type M association should be defined from STATE to CITY. The designer may want to include

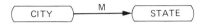

Figure 4-11. Type M association.

Figure 4-12. M:1 mapping.

an inverse Type 1 association to specify that the given city belongs to only one state (Figure 4-12).

In some situations, the designer may inadvertently use a Type 1 association from a source key to a target key in an incorrect manner. The structuring algorithms always interpret a Type 1 association between keys as defining a child-to-parent relationship in which the source ("from") key represents the child and the target ("to") key represents the parent (i.e., a child has one and only one physical parent). But sometimes the designer wants the target of a Type 1 association to be a child segment having only one occurrence. An example from the insurance industry is in defining policy number as a root segment and policy description as a dependent child segment having only one occurrence. The view of Figure 4-13 will not produce this result because the Type 1 association from POLICY-NUMBER to DESC-ID will be interpreted as defining a child-to-parent relationship rather than the expected parent-to-child relationship.

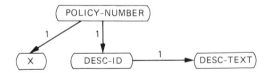

Figure 4-13. Type 1 association to DESC-ID produces undesired result.

Although the Type 1 association defines a unique identification, it does not cause a target key to be structured into a role subservient to the source key. In a binary relationship between keys, if the target ("to") element represents data that is to be subservient to the source ("from") element, the association from source to target should be Type M or Type C to obtain the desired parent-to-child structure (Figure 4-14). Remember that the Type M association represents identification of zero, one, or more than one occurrence of the target element; the Type C association specifies unique identification if the target element exists.

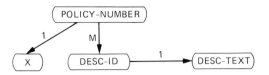

Figure 4-14. Type M association to DESC-ID produces desired result.

Content of Local Views

The content of the local views submitted to the automated procedures is defined below. Of the items to be specified, some are required for logical structuring, while others are optional. The amount of design information provided by the computer-assisted process is increased as more optional items are specified.

An interactive language for entering this information into machine-readable form for the automated procedures is suggested in Chapter 22.

Required Information

The following items are required for any use of the automated procedures:

- Name of the local view
- Names of the data elements
- Association types (showing the element pairs being related)

With this information, the procedures to be described can perform most of their editing and can derive a basic logical model consisting of segment contents and physical parent-child relationships. However, the resulting logical model may contain unresolved alternatives that could have been resolved automatically had more input information been available.

Optional Information

The following items of optional information will enrich the results of the automated procedures:

- Frequency of use and setup period of the function represented by the local view.
- Estimated accessing frequencies and processing options of the data elements in the local view.
  The information in these two items can be used in calculating perfor-

mance weights from which several performance-related decisions can be made. The calculation and use of performance weights are discussed in Chapter 10.

- Expected number of occurrences of each data element.

  To be used in estimating data base size and in judging performance implications of searching twin chains.

- Data element characteristics (e.g., length, type, and format).

  To be used for calculating segment sizes and for editing the use of the same data names from different local views.

- Processing option for each use of a data element.

  To be used for calculating performance weights (see Chapter 10).

- Data element sources.

  To indicate that there is a source, either as input data or as the result of a calculation, for all data used by a local view.

SPECIAL DATA GATHERING AND RECORDING CONSIDERATIONS

In the process of gathering the data requirements, certain situations may require special consideration. Either it is not obvious how these situations should be expressed in bubble charts or it is awkward to do so. Some of the more commonly encountered situations of this type will be discussed here.

It is unnecessary for these situations to be recognized and specified in the initial gathering of the data requirements. Indeed, some of them are not likely to become evident until after the several local views have been merged into a composite model or until a logical design has been derived. The automated (or manual) procedures should be able to accept specifications of these situations in any iteration of the design procedures.

An important concept that bears repeating at this point is that when providing the initial data requirements, the designer's (or end user's) thinking process should be free of structural considerations so that the specified requirements reflect the application function's true needs in as unbiased a manner as possible. In practice, the designer and the end user will not be able to avoid thinking of structural implications as the data requirements are being specified. In such situations, their thinking should at least be limited to their own local view. Although structurally oriented decisions and choices can be included at any time, they are normally more appropriate after the first iteration.

It is premature to discuss structuring implications at this point because the methods for deriving structures from binary relations and their association types have not yet been discussed. But some of these implications are useful at this point to help the designer or end users specify the initial data requirements in a proper manner. The reader may want to review this section again after studying the chapters on logical design.

Compound Keys

Frequently, more than one source data element is required for identification of a target data element. For example, coaches keep records of the performance (number of tackles, number of passes caught, number of yards gained, etc.) of each player in each game. PLAYER-NUMBER and GAME-NUMBER are both required to identify a given performance statistic; PLAYER-NUM-BER*GAME-NUMBER is a compound key.

Compound Keys are combinations of two or more data elements, all of which are required to identify a target element. Both CLASS and STUDENT are normally required to identify GRADE. A particular grade pertains to a particular student in a particular class. Therefore, CLASS*STUDENT is a compound key of GRADE.

Compound keys can be represented in the bubble charts shown in Figure 4-15.

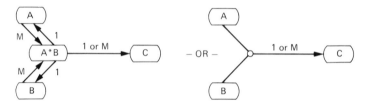

Figure 4-15. Specifying compound keys.

In the right-hand notation, which is simpler to draw, the "o" represents the compound key A*B; the M:1 mappings between A and A*B and between B and A*B are implied. The meaning of the bubble chart is that an instance of A and an instance of B are both required to identify an instance of C. The M:1 mappings between each of the simple keys (A and B) and the compound key (A*B) are inherent to the concept of compound keys. For example, for a given student in a given class (STUDENT*CLASS), we have uniquely identified the student. But we can have many occurrences of STUDENT*CLASS for a given student because the student may be enrolled in many classes. In DL/I structures, the compound key A*B defines an additional segment that may serve as an intersection between segments A and B.

Roles

We have seen that the relationship from PLAYER-NO to POSITION can be interpreted in more than one sense or Role. In one role POSITION can be the position where the player is currently playing and in another role it represents all the positions at which the player can play. Consider an association from

CUSTOMER to ADDRESS. In one role ADDRESS can be the location to which the customer's bill is to be sent, and in another role it can represent a different location to which the order is to be sent.

In CODASYL structures, different roles of the target element are explicitly indicated by labels assigned to the association. In DL/I and relational structures, roles are implicitly indicated. There are several ways of making this indication for DL/I structures and each has its structural implications. Three possibilities will be explored below. After examining the results of a logical design iteration, the designer may want to respecify certain roles in order to produce a more desirable structural result.

One common way of indicating role is to select more than one attribute from a common domain and give each such attribute a uniquely qualified name (Figure 4-16). This results in the creation of redundant data which could have adverse storage space and update integrity implications.

Figure 4-16. Specifying roles (1).

The structural implications of Figure 4-16 are that the target elements of the Type 1 associations, unless they also are keys, will be structured as attributes into the segments of their keys (PLAYER-NUMBER or CUSTOMER in this case). Unless additional associations are present (perhaps from other local views) to add further detail, target elements of Type M associations, as shown in Figure 4-16, are classified as Floating Elements because they can be structured in either of two ways: (1) as variable length repeating fields also in the same segments as their keys or (2) as dependent segments to their keys. The discussion of repeating attributes in Chapter 8 gives further details.

For Type M associations to the target element, another way of indicating roles for DL/I structures is to create an additional attribute (e.g., flag field) indicating the type of role (Figure 4-17). Using the player-position example,

Figure 4-17. Specifying roles (2).

the result will be a single set of positions per player, and they will be structured into a dependent segment type with TYPE as one of its fields. The application program would have to interrogate TYPE to determine the role of POSITION or ADDRESS. Redundancy is not removed because the positions that apply to one player will also apply to other players. Therefore, the same position data may be replicated across several players.

A third way of specifying role is by defining special elements that will serve as intersections between player and position and which will be structured as logical children. This method may be used when it is desired to avoid redundant storage of data. In IMS this approach can be implemented by logical relations in which the special elements are used to define intersections between PLAYER-NO and POSITION (Figure 4-18). Redundancy problems are avoided and program interrogation is not required to determine the role. Logical relations can be slower to search than physical parent-child relationships, but they also provide convenient bidirectional capabilities.

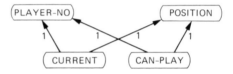

Figure 4-18. Specifying roles (3).

Cyclic Relationships

Cyclic relationships are a special case of roles and exist when one occurrence of a data element refers to other occurrences of the same data element type. An example is the bill of materials application (e.g., parts, explosions, and where-used lists) in which parts are components or assemblies of other parts.

Cyclic relationships can be specified as shown in Figure 4-19.

The first case expresses a relationship between A and its superset which identifies the assemblies containing A; the second case expresses a relationship between A and its subset which identifies the components of A. The automated procedures should be able to recognize the A*SUP and A*SUB as special role

Figure 4-19. Specifying cyclic relationships.

identifiers of A, to automatically structure pointer (intersection) segments, and to make appropriate entries in the list of candidates for logical relations to be produced in the logical design phase.

Entity Subclasses

Some entities may be defined in terms of their subclasses. For example, managers are also employees; therefore, managers may be defined as a subclass of employees and there may be attributes for managers that are not specified for employees in general. Another example deals with locations. In the petroleum industry, a location may be a service station, a refinery, or a distribution center. There may be attributes for location in general and additional attributes for each type of location.

Data about subclasses are usually implemented in one of three ways: (1) as additional attributes in the entity's segment and with a flag field to indicate which type of subclass is being described, (2) as a single dependent segment type containing attributes for the subclass types and with an identifying flag field, and (3) as a separate dependent segment type for each subclass type, only one of which will exist for each entity.

Bubble charts that will lead to these three implementations are illustrated in Figures 4-20a, 4-20b, and 4-20c, respectively.

In Figure 4-20a, A, B, and C represent those attributes common to all employees, and X and Y represent those attributes unique to managers. In Figure 4-20b, A, B, and C represent general attributes for all locations inde-

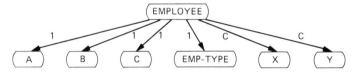

Figure 4.20a. Specifying entity subclasses (1).

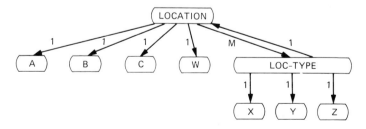

Figure 4-20b. Specifying entity subclasses (2).

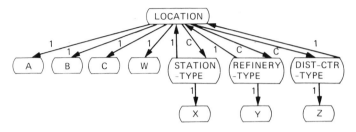

Figure 4-20c. Specifying entity subclasses (3).

pendently of subclass. X, Y, and Z represent attributes unique to each subclass type. W represents additional subclass information common to all subclasses. In Figure 4-20c, the attributes keep these same relative meanings.

Finally, the end user may want to refer to an entity by a unique subclass name rather than by a general entity name. A special SUBCLASS editing command described in Chapter 5 can provide this capability.

Summary Data

Summary data are typified by totals, averages, and other statistical summaries. Whether or not summary data should be stored in the data base or recalculated as needed depends on their frequency of use, storage space requirements, and retention requirements. Not infrequently, summary data are to be retained after the detail from which they are calculated is deleted.

When stored, summary data should be distinguished from their detail data by name and by other role considerations where applicable. An example of this practice is illustrated in Figure 4-21.

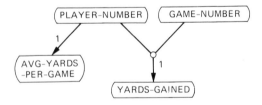

Figure 4-21. Specifying summary data.

Sort Sequences

Often several reports having the same content but different sort sequences are required. For example, consider a reporting system for a football team in which the Player Roster contains the number, name, position, height, and experience of the players. This Player Roster may be ordered on number, on name, or on any of the data elements it contains, depending on the purpose of the report. Figures 4-22a, 4-22b, and 4-22c illustrate how various sorting requirements may be represented in bubble charts.

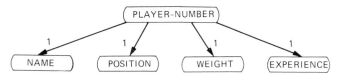

Figure 4-22a. Specifying sort sequence by PLAYER-NUMBER.

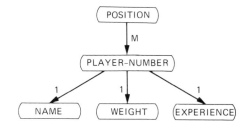

Figure 4-22b. Specifying sort sequence by POSITION.

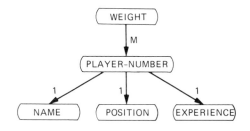

Figure 4-22c. Specifying sort sequence by WEIGHT.

Assume that PLAYER-NUMBER uniquely identifies each of the other data elements. Assume also that there can be several players for each position and several players of a given weight. If reports are to be ordered on PLAYER-NUMBER, POSITION, and WEIGHT, the three bubble charts of Figure 4-22 could be drawn.

Figure 4-23 shows the composite model that will be deduced from these three requirements.

The automated procedures can derive a logical design for these requirements. It will be seen later that if certain performance-related information can

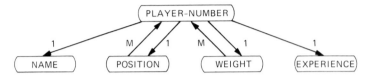

Figure 4-23. Composite model of sort sequences.

be included with the data requirements, the procedures can also calculate performance weights to help derive the most efficient storage structure and to help the designer determine to what extent secondary indexing or sorting might be warranted.

Special Editing Commands

Some special editing commands, suggested in Chapter 5, could also be used by the designers in recording the initial data requirements if they have sufficient insight into the nature of the data and its relationships. These commands are concerned with resolving such things as synonyms, data groups, and entity subclasses and with specifying data element characteristics such as length, type, and so forth. They are intended primarily for specifying editing and structuring preferences for iterations after the first.

PRAGMATIC DATA ENTRY CONSIDERATIONS

Who Should Provide the Data Requirements?

One opinion is that the data requirements should be recorded initially by the end users themselves. Thus, the owner of each local view provides his own requirements directly to the automated procedures. An opposing viewpoint is that the designer(s), functioning as the data base administrator(s) in dialog with the end users, should specify the initial requirements.

The advantage claimed for the former approach is that the initial local views will more accurately reflect the user's own perceived views of his data requirements and will be less biased by preconceived data structuring considerations. On the other hand, the designer, in close contact with the end user, will usually record local views that are more consistent with one another and with the rules of the data base management system to be used. The designer will tend to preedit and prestructure the data requirements, thus performing, to a degree, some of the work of the automated procedures. Requirements provided directly by the end users may be more pure in terms of their functional requirements, but will probably contain more inconsistencies and require more editing iterations than if the designer assists in specifying them.

Another consideration is that if the task of initial data entry is also divided among the several end users, the large size of the task becomes much less formidable than if one person (or one group) were doing it all. And one or more retranscriptions of the data requirements may be avoided.

Probably the most important consideration is that the end users should receive a feeling of confidence in the designers and in the integrated data base approach as early in the process as possible. They should receive positive results, as soon as possible, that editing is being done that is helping them to define their data requirements more clearly and consistently. Regardless of which approach is used, experience indicates that the more the end users are involved in the data gathering and recording process, the more confidence they will have in the design study as a whole, and the more cooperative they will be with the designers and with the data processing personnel.

Reducing the Labor of Data Entry

Describing local views in the form of binary relations and bubble charts or both, and preparing this information for machine or human processing, can be a gigantic task. In a large design study there can be hundreds of local views and thousands of data elements. Many of the data elements will reoccur in several local views. Thus, the process of recording the data elements and their associations in machine-readable form can be monumental in terms of time and tedium. In a well executed design study, this is work that has to be done anyway, in some form, whether or not computer-aided techniques are used. The question, then, is not should the work be done, but rather: How can the impact of its magnitude and tedium be minimized?

The data requirements should be entered only once into machine-readable form, not into a dictionary system, and separately into the design procedures. Preferably, they will be entered into a dictionary system that can provide input to the automated design procedures in addition to its other functions. An added benefit is possible if the design procedures and the dictionary system are designed so that many of the editing procedures of the conceptual design phase can also filter the data descriptions to be permanently stored in the dictionary. Desirable interfaces between the automated design procedures and a dictionary system are explored in Chapter 21.

With a dictionary system, full descriptive information about the data elements will normally be entered. Each data element will be represented, some of them more than once when they occur in different local views. Manual attempts to reduce this duplication before data entry are usually less desirable than entering every occurrence of each element and letting the dictionary and the design procedures or both perform automated editing.

Without a dictionary system, there are several ways of reducing the amount of data entry, but each has its drawbacks. The only information absolutely required by the structuring procedures are the names of the local views and the names and relationships of the data elements in each local view. But it was pointed out earlier in this chapter that a large degree of editing, performance information, and resolution of alternatives will be sacrificed by minimal data entry. And if going from automated logical design into automated physical design, the information from all these features will be vital.

Another labor-saving approach, if considering logical design only, is to enter only the binary relationships between keys of entities already identified. This approach, however, presupposes a portion of the logical design and, without the attributes, the resulting editing and diagnostic reports will lose significant richness.

A final pragmatic consideration involves reducing the number of times the data requirements are transcribed on paper before being entered into the system. It is likely that the designers or end users will record the local views by drawing bubble charts or by some other equivalent method. This should be done so that it does not need to be rerecorded for machine entry. Interactive input of the data requirements by using a keyboard display terminal is appropriate for this type of data entry. For each local view, the end user usually carries in his head, or has on paper close at hand, his perceived descriptions and other characteristics of the data elements which he could enter directly. Otherwise, it will probably have to be specially recorded for entry by the designer and his staff.

A combination of well-designed menus and formatted screens can guide the terminal user through the entry process in a natural way that minimizes keying. A sample data entry scenario showing a possible use of screens and menus is included in Chapter 22. To the extent that interactive data entry into an appropriate dictionary system is available, it should be used instead of separate data entry procedures.

5. The Editing Process

Obtaining an understanding of the end users' true requirements and establishing standard usage, meanings, and names of the data are some of the most crucial aspects of data base design. While these are primarily human functions, automated analysis and reporting can be very helpful. The purpose of the editing phase, then, is (1) to help promote a consistent and error-free gathering and recording of the data requirements, (2) to assist the designer in policing data usage and naming standards, and (3) to produce a composite model of the data requirements that truly supports the perceived view of the enterprise.

We will first suggest some editing reports that can be produced, and then we will discuss the three editing objectives in terms of those reports. Finally, we will suggest some special editing commands that can assist the designer in specifying the resolution of certain editing problems.

EDITING REPORTS

Automated analysis of the data requirements can produce a number of reports that can be very helpful in revealing inconsistencies in the use and naming of the data. Because only humans can associate semantic meaning with the names of data elements, automated editing procedures are not able to reveal all the inconsistencies and problems that may exist. But these procedures can produce a variety of reports that can give significant assistance to the person who is editing. Well-designed editing reports can form the basis of much of the dialogue that must take place between the designer and the end user in order to assure data requirements that are complete, consistent, and meaningful.

Instead of presenting possible formats for these reports, their desired contents will merely be indicated. Since initial data entry may be made by either the designer or by the end user, we will use the term *user* in a generic sense in most of this section.

Edited Local Views

The local views, as entered, can be displayed back to the user with syntactic errors noted and diagnostic comments included. "Best guess" resolutions of many of the errors can also be supplied, and this is recommended if the ability for user review and overrides is also provided.

Data Definition and Where-Used List

An alphabetical list of all data element names can be produced; this names all the local views in which each name is specified and lists all the characteristics and descriptions available about each name. Inconsistencies in the characteristics and descriptions can be detected and reported. The list can be updated automatically as inconsistencies are resolved and as new local views are added into the design study. In addition to a where-used list for individual data names, a where-used list for each association between related data name pairs can also be useful.

Keyword-In-Context (KWIC) List

A list showing the contexts in which all the qualifications of the data names are used can be very helpful in looking for synonyms, homonyms, and other naming inconsistencies. Consider the following example (Figure 5-1) from local views for the football league data base.

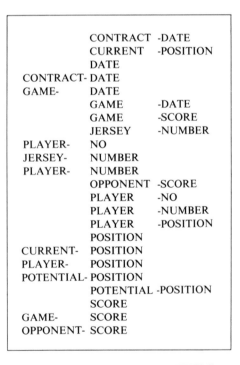

```
                    CONTRACT  -DATE
                    CURRENT   -POSITION
                    DATE
        CONTRACT-   DATE
        GAME-       DATE
                    GAME       -DATE
                    GAME       -SCORE
                    JERSEY     -NUMBER
        PLAYER-     NO
        JERSEY-     NUMBER
        PLAYER-     NUMBER
                    OPPONENT  -SCORE
                    PLAYER     -NO
                    PLAYER     -NUMBER
                    PLAYER     -POSITION
                    POSITION
        CURRENT-    POSITION
        PLAYER-     POSITION
        POTENTIAL-  POSITION
                    POTENTIAL -POSITION
                    SCORE
        GAME-       SCORE
        OPPONENT-   SCORE
```

Figure 5-1. Keyword-in-Context (KWIC) list.

Looking at this KWIC list, the designer can see many things that should be questioned. For example, there are four different names containing POSITION. While CURRENT-POSITION and POTENTIAL-POSITION may be clear, does PLAYER-POSITION refer to one of these or to something different? And what about POSITION without any qualification? Ambiguities and possible homonyms appear to exist. Synonyms appear to be present in JERSEY-NUMBER, PLAYER-NUMBER, and PLAYER-NO. Should GAME-SCORE be a group name for two fields, one of which might be TEAM-SCORE and the other OPPONENT-SCORE? A number of other questions can also be asked about this example.

Inconsistent Associations

Inconsistent Associations exist when two associations of different types are specified in the same direction between a pair of elements. Consider the examples in Figure 5-2.

Figure 5-2. Direct inconsistent associations.

Homonyms and the need to assign roles or simply reevaluate the data requirements are suggested by such a presentation. For example, in the football league, one user may be thinking of the unique position to which a player is currently assigned, while a different user is thinking of all the positions at which the player can play. In the insurance case, one user may be thinking of the unique person covered by one type of policy, while another user may be thinking of the group of persons covered by another type of policy. Or, in another sense, one user may be thinking of the single person to be billed by a policy, while another user may be thinking of the group of persons covered by the same policy.

Direct instances of inconsistent associations, as illustrated in Figure 5-2, are relatively easy to detect with computer assistance. Detecting indirect inconsistent associations, which are illustrated abstractly in Figure 5-3, is more difficult. These should also be identified and reported.

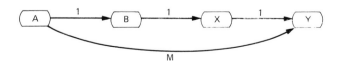

Figure 5-3. Indirect inconsistent association.

The early finding of all instances of inconsistent associations by manual techniques is very unlikely when designing a large data base from many local views.

Intersecting Attributes

This is a list of all attributes having more than one key. Although these attributes and their keys may represent true redundant data, the multiple keys may also be the result of naming inconsistencies or the need to assign roles. Identification of redundant data is also discussed in the logical design process.

Figure 5-4. Intersecting attributes.

In the first example of Figure 5-4, VENDOR and SUPPLIER are probable synonyms that would not have shown up as such on the other editing reports. VENDOR may have been used in one local view and SUPPLIER in a different local view. In the second example, PHONE should probably be given a qualified name and made an attribute of each of its keys.

New Element Report

When new functions are to be added to an existing data base design, many of the required data elements and relationships are likely to be a part of that design already. With very little additional effort, the editing process can provide a list of data elements and associations that are new to assist the designer in evaluating possible trade-offs in case of design conflicts.

Source-Use Exceptions

Except to provide for anticipated future requirements, every data element should be used as output or in a calculation and should be provided by an input or by a calculation. A list of exceptions will assist the designer in judging the "clutter" and the availability of the data.

OBJECTIVES OF EDITING PHASE

The following objectives are proposed for the editing phase.

Consistent and Error-Free Input

To prepare input for systematic procedures, and especially for those automated as computer programs, a high degree of completeness and consistency is required. In fact, one of the major benefits of using computer-assisted procedures at this point is that they force a more thorough analysis of the functional and intrinsic data requirements than might otherwise be done. Extra effort at this stage of the design cycle can pay significant dividends in shortening the time of the overall design cycle.

The local views as entered must be specified to a consistent level of detail. They can be edited for syntax errors, according to the rules and expectations of the automated procedures, with the edited local views and diagnostics being returned to the designer. In many cases, areas that are incompletely specified and that are structurally ambiguous can be detected by using the editing reports.

Naming Standards and Usage

The various possible editing reports can provide significant assistance in detecting problems and inconsistencies in the usage, format, and meaning of the data elements specified in the various local views. An alphabetical list of data names from the Data Definition and Where-Used report, the KWIC list of names and their qualifiers, and an Intersecting Attributes list can all help the designer in finding synonyms and homonyms. The Inconsistent Association report may also reveal homonyms or may indicate where roles should be assigned or simply where requirements should be reconsidered. The Edited Local Views, as well as the Data Definition and Where-Used report, can help reveal different usages, formats, and meanings expected of the data elements.

Once naming standards are established, an automated editing procedure can check names submitted in local views against a list of standardized names and can report discrepancies back to the designer. Again, the concept of an iterative editing phase is suggested; the automation makes the iterations much more convenient to perform.

Representative Composite Model

The editing phase by itself is an iterative process to be repeated until all diagnostics are resolved. In consultation with the end users, the designers insert changes and reinterpretations into the local views, and repeat this process until, hopefully, all parties are satisfied. The end result is a composite model that represents the requirements of the edited local views and from which logical and physical models can be derived.

SPECIAL EDITING COMMANDS

The following are some Special Editing Commands that are suggested for the automated process. They can provide a convenient means of injecting the resolution of certain situations into the primary input. Again, it it necessary to allude to the structural implications of these commands and the reader may want to review this section after reading Part III of this book.

Synonyms

In standardizing data element names, synonyms must be resolved by instructing the editing procedures that two or more names refer to the same data element. Rather than correct every instance where these names may appear in the data requirements, a single SYNONYM command is desirable. Given that A and B are two names which the designers determine to be synonyms, the SYNONYM command would specify which name is to be retained, and the associations would be adjusted accordingly. For example, if the example of Figure 5-5a had been specified, the SYNONYM command should produce the interpretation shown in Figure 5-5b.

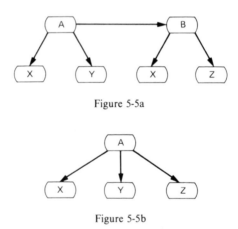

Figure 5-5a

Figure 5-5b

Figure 5-5. Resolution of synonyms by SYNONYM command.

Entity Subclasses

It may be desirable to refer to a particular entity subclass by a unique name rather than by the general name of the entity. A SUBCLASS command can be used to specify these unique subclass names and relate them to the general entity name. This command could also be used to specify entity attributes that are unique to the subclass and the type of structuring desired.

Groups

Several data names may represent subelements of a more generic grouping. For example, DAY, MONTH, and YEAR may represent subelements of DATE. If these subelements will always be stored together as concatenated values in the same field, they can be represented by a single group name (e.g., DATE). The procedures should provide a GROUP command to permit the assignment of group names. During editing, the individual names should be used as specified, but after the editing phase is completed the automated procedures should work only with the group name.

Assigning group names on a global basis will sometimes add confusion to requirements from some local views. The designer should decide if renaming certain elements should precede use of the GROUP command.

Characteristics Definitions

When a data element is specified on different local views, it is not uncommon for it to be specified with different characteristics such as length, format, type, or description. After an appropriate dialogue between the designers and end users, a DEFINE command will be useful for specifying the "official" characteristics which will override the others already specified in various local views.

PART III
LOGICAL DESIGN

The logical model of a DL/I data base is defined to be the collection of data elements into segments and the collection of segments into physical hierarchies with logical relationships and secondary indexes indicated at appropriate places. The physical storage of the data—including device type, access methods, block lengths, grouping, pointer options, insert-replace-delete rules, etc.—are not considered in the logical model. These physical considerations are ingredients of physical design which follows logical design.

In this section we present the computer-assisted procedures for deriving a logical model from the composite model constructed in the conceptual design. Primary emphasis is on designing for DL/I structures.

Chapters 6 and 7 describe the procedures for deriving the physical aspects and the logical aspects of the logical model. These procedures are straightforward; however, some subtle results or anomalies can be obtained, and these are reviewed in Chapter 8. In following these procedures, several areas can be encountered where design alternatives exist. Chapter 9 explores the criteria for making some "best guesses" in these areas. In addition, while deriving the logical model, certain performance-oriented information can also be obtained as described in Chapter 10.

One of the primary benefits of computer-assisted design is the wealth of diagnostic information that can be made available to the designer early in the design cycle. Possible diagnostics are outlined in Chapter 11 along with a review of desirable reports of the logical design. The results of any automated design procedure should be carefully reviewed by the designer. Chapter 12 emphasizes that computer-assisted design should be an iterative process in which the designer, using the diagnostics provided as well as his own design intuition, refines the results of the previous iteration until he converges onto the final design to be implemented.

Finally, in Chapters 13 and 14, concepts for extending computer-assisted logical design to relational and CODASYL structures are explored.

6. Logical Structuring: Physical Relationships

PRELIMINARY CONCEPTS

Before presenting the procedural steps for structuring the logical model, certain preliminary concepts should be established. These concepts deal with the derivation of segments and physical parent-child relationships and are based on the structural implications of the definitions that have been established for keys and attributes.

Keys, Attributes, and Segments

As stated in Chapter 2, any element that identifies any other element with a Type 1 (simple) association, is (by definition) a key. In DL/I, a key is the sequence field of a segment. Thus, when a data element is classed as a key by the procedure to be outlined, it automatically defines a DL/I segment.

An attribute is any nonkey element identified by a Type 1 (simple) association. Thus, every attribute has a key and, in the initial logical model, each attribute will be clustered into its key's segment as shown in Figure 6-1.

Physical Parent-Child Relationships

Between keys, there are only three mappings that can yield a physical parent-to-child relationship. Assume all elements are keys in Figure 6-2.

The example of Figure 6-2a represents the classic definition of the physical parent-child relationship in which each parent occurrence may have any number (zero, one, or more than one) of occurrences of a child segment type, while each child segment occurrence pertains to one and only one physical parent occurrence. The other two cases are special cases of the first. Figure 6-2b represents a parent that may or may not have an occurrence of its child, but if it has, it has only one occurrence. In Figure 6-2c a parent occurrence always has one and only one occurrence of its child.

These three key-to-key mappings have a common characteristic: the Type 1 association from child upward to parent. The association from parent downward to child is different in each case. Thus, the structuring procedures to be outlined will rely on the Type 1 associations between keys as the defining factor for determining physical parent-child relationships. These procedures will build

Figure 6-1. Derivation of segments.

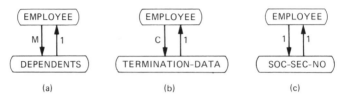

Figure 6-2. Mappings for physical parent-child relationships.

the physical hierarchies in a bottom-up manner by following Type 1 associations from key to key—lowest key to highest (root) key. Because the case of Figure 6-2c has Type 1 associations in both directions, an anomaly exists that will be treated in Chapter 9.

Structuring from the Composite Model

Each designer should understand that the local views he submits may not be the sole criteria for classifying the data elements. If some of these elements are also named in another function's local views (a thing to be expected, and even encouraged, in designing integrated data bases), such elements may be classified differently from what is expected.

For example, if Designer 1 submits the binary relation of Figure 6-3a as part of a local model, a segment with A as a key and X as an attribute might be

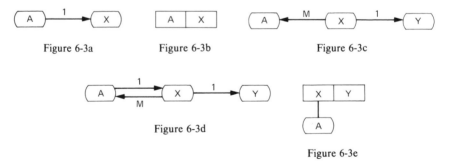

Figure 6-3. Concept of structuring from the composite model.

expected (Figure 6-3b). But if Designer 2 submits the requirement of Figure 6-3c into the same design study, the resulting composite view will be that of Figure 6-3d which places X into its own segment as a parent of A (Figure 6-3e). The functional requirements of both users are satisfied, but perhaps not as expected and they will want to review the performance implications of the resulting structure. It will be seen later that A, in this case, may be listed as a candidate for secondary indexing.

THE STRUCTURING PROCESS

With the basic structuring concepts in mind, the procedural steps for deriving a logical design can now be presented. This process for structuring a composite model into a DL/I logical model is outlined in the remainder of this chapter and in Chapter 7. In succeeding chapters, subtleties and alternatives resulting from this process are reviewed.

The basic structuring process consists of the following seven steps:

1. Determine keys and attributes
2. Augment Type M associations between keys
3. Remove transitivity
4. Construct suggested segments
5. Build physical hierarchies
6. Determine candidates for logical relationships
7. Determine candidates for secondary indexing

The last two of these steps form the content of Chapter 7.

Determine Keys and Attributes

As has been stated, those elements from which a Type 1 association emanates are defined to be keys. Nonkey elements which are targets of Type 1 associations are attributes. The keys may be simple or compound as illustrated in Figure 6-4.

Augment Type M Associations Between Keys

The structural implication of a Type M association between keys is not fully defined unless an inverse association has also been specified between the two keys. An association for which no inverse has been specified is called a Lone Association. Lone Type M associations (Figure 6-5a) represent relationships that can be implemented in either of two ways (Figure 6-5). They can be implemented as physical parent-child relationships (Figure 6-5b), or they can

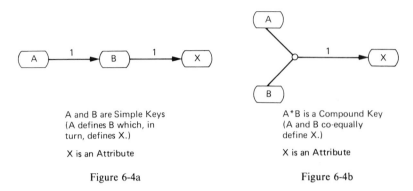

A and B are Simple Keys
(A defines B which, in
turn, defines X.)

X is an Attribute

Figure 6-4a

A*B is a Compound Key
(A and B co-equally
define X.)

X is an Attribute

Figure 6-4b

Figure 6-4. Deriving keys and attributes.

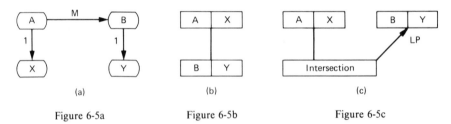

(a)

(b)

(c)

Figure 6-5a

Figure 6-5b

Figure 6-5c

Figure 6-5. Interpreting Type M associations.

also be implemented in physical parent-logical parent relationships (Figure 6-5c), provided a suitable intersection is also specified.

On the other hand, natural design thinking often occurs in a top-down sense and is described in terms of lone Type M associations. For example, a designer may think of a division as identifying many departments, of each department as identifying many employees, and of each employee as working on many projects; from this, he could draw the bubble chart of Figure 6-6.

To require the designer to specifically supply an inverse association for every instance of a Type M association would mean a significant increase in the time and labor of specifying the data requirements, and it should be avoided except where necessary. Thus, the following procedure can be used.

Unless there is an explicit need for the logical relationship (as defined by an M:M mapping), the physical parent-child structuring is usually preferable for simplicity and for performance. Therefore, as an optional feature after the local views have been merged, the automated procedures will supply, by default, a Type 1 inverse (Figure 6-7a) to define physical parent-child structuring (Figure 6-7b) for Type M associations between keys that still have no inverse. The alternative is to allow the procedures to assume logical relationships for all lone Type M associations between keys.

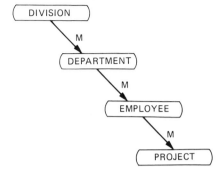

Figure 6-6. Specifying lone Type M associations.

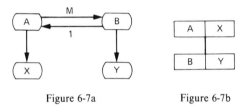

Figure 6-7a   Figure 6-7b

Figure 6-7. Augmenting lone Type M associations between keys.

The designer should receive a list of Type 1 associations supplied in this manner so he can determine if the assumed structuring satisfies his requirements. And he should explicitly specify the M:M mapping (Type M associations in both directions) when he requires logical relationships rather than the physical parent-child relationship.

Remove Transitivity

A Type 1 association between two data elements is said to be Nonessential (or Transitive) if there is at least one other path of Type 1 associations leading from the source element to the target element. Presumably, the transitive association can be removed without destroying the capability of traversing from the source to the target element. Figure 6-8 illustrates the transitivity property. By removing the association from A to C, one can still traverse from A to C via the intermediate element B.

Transitivity among Type M associations is more elusive, but it does not concern us in these procedures because it is the Type 1 associations that are being analyzed to determine the physical relationships between data elements (i.e., keys, attributes, and physical parent-child relationships). The Type M associ-

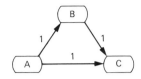

Figure 6-8. Concept of transitivity.

ations will be analyzed only to determine candidates for logical relationships and secondary indexes.

The purpose of removing transitivity is to eliminate potential redundancies and potential update anomalies as explained by Date in Reference A.4. But removal of transitivity can also produce undesirable design results in DL/I structures. Although access can still be made from A to C by going from A to B and then from B to C (Figure 6-8), the following questions become pertinent to the designer:

- Can the application stand the performance degradation of extra accesses to reach the target element?
- Will the desired occurrence (value) of the target element be obtained by taking a different route?
- Will there be a loss of information?

As examples of these last two questions, consider the following cases. Here we are really dealing with "apparant" transitivity, while the real problem is usually one of improper naming of data elements.

Case 1

Removal of the transitive association from EMPLOYEE to PHONE in Figure 6-9a could cause queries for employee phone number to return the manager's phone number. A likely solution (Figure 6-9b) is to define separate phone number elements, with distinguishing qualified names, as attributes of employee and of manager.

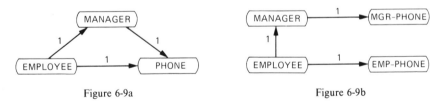

Figure 6-9a   Figure 6-9b

Figure 6-9. Removing transitivity possible problems (1).

Case 2

Removal of the transitive association from PLAYER to NO-PENALTIES in Figure 6-10a will result in loss of the information about the number of penalties assessed against each player in a game and will leave only the total number of penalties assessed. One possible solution (Figure 6-10b) is to retain the association from PLAYER to NO-PENALTIES, remove the association from GAME to NO-PENALTIES, and let the application calculate the total penalties per game by accessing the number of penalties for each player in the game. Another solution (similar to Case 1) is to maintain the total number of penalties as a separate element of summary data (Figure 6-10c). The proper choice depends primarily on the frequency of use of these paths for inquiry or update and on the amount of additional searching required with each use.

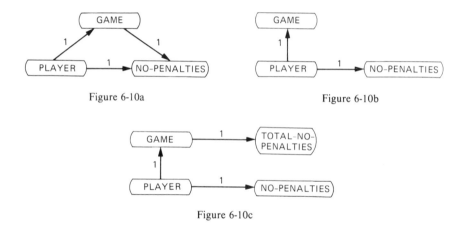

Figure 6-10a

Figure 6-10b

Figure 6-10c

Figure 6-10. Removing transitivity possible problems (2).

The designer should receive a list of transitive (nonessential) associations that have been identified (and thus removed), and he should be given the capability of causing any of them to be reinserted into the design. This can be by direct reinsertion or by changing data element names as illustrated above. Many instances of transitivity can be avoided by adopting appropriate naming standards and practices.

A programming procedure for detecting and removing transitivity is given in Reference C.2.

Segments derived after removing transitivity are said to be in third-normal form becuase the following conditions exist:

- All elements are atomic (single-valued) (first-normal form).

- All attributes are identified by a single (primary) key, which may be a compound key (second-normal form).
- No attribute is identified by any other attribute in the segment (third-normal form).

The concept of deriving third-normal form structures is explored more fully in Chapter 13.

Construct Suggested Segments

Each key, by definition, defines a segment, and the attributes clustered about a key are mapped into its segment (Figure 6-11). The third-normal form segments automatically constructed in this manner (after transitivity has been removed) should be considered as suggested segments subject to revision by the designer. It will be suggested in Chapter 12 that the designer may want to split such segments having too many fields or combine segments having few fields. He may want to group fixed length fields into fixed length segments and variable length fields into variable length segments. Or he may want to categorize and separate fields according to accessing frequencies or security constraints.

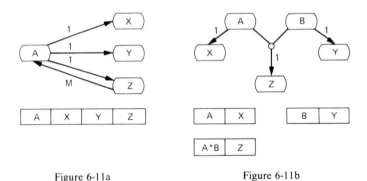

Figure 6-11a   Figure 6-11b

Figure 6-11. Constructing suggested segments.

Build Physical Hierarchies

Physical hierarchies are built by tracing paths of Type 1 associations upward from child to parent. Starting with each Low-Level Key (i.e., a key that does not have another key pointing to it with a Type 1 association), the path of Type 1 associations is followed until a high-level key is encountered. A High-Level Key is one that does not point to another key. Several paths may lead to the same high-level key. The high-level key defines the root segment of the resulting hierarchy. The low-level keys leading to it define its low-level segments. And the other keys encountered in the several paths define the intermediate-level segments.

Consider the composite model of Figure 6-12a. A, B, C, and D are keys and therefore define segments. C and D are low-level keys, and A is the only high-level key. Following the Type 1 associations from low-level keys to the high-level key yields the physical hierarchy of Figure 6-12b.

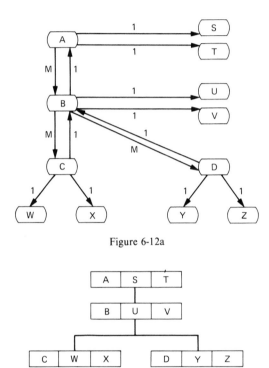

Figure 6-12a

Figure 6-12b

Figure 6-12. Building physical hierarchies.

In tracing the paths of Type 1 associations, two violations of DL/I rules (Reference D.5) may be encountered. The path may loop back onto itself, or the path may contain more than 15 keys. Either of these conditions will prevent correct DL/I structuring. Keys involved in a loop cannot be structured auto-matically into a parent-child relationship without arbitrarily breaking the loop at some point. Therefore, loops must be reported as a diagnostic. Paths with too many levels can be structured, but the lower levels (beyond 15) could be dropped and reported also as a diagnostic. Both situations require the designer's resolution.

In addition, segments having more than one candidate parent can be detected and should be reported to the designer. If performance weights, which will be discussed in Chapter 10, are present to indicate a relative importance of the hierarchical paths, the parent on the most important path can be automatically chosen as the physical parent. If there is only one additional parent, it can be implemented as a logical parent. If there is more than one additional parent, the designer must resolve the issue. In all cases, the designer must have the ability to override the procedure's choice of physical parent. The important point is that all instances of this type can be recognized and reported to the designer.

Finally, the reader should notice that in determining segments (i.e., keys and attributes) and physical hierarchical structures, only the Type 1 associations have been considered. The Type M associations will be used in Chapter 7 for determining candidates for logical relationships and for secondary indexes.

7. Logical Structuring: Logical Relationships

CANDIDATE LOGICAL RELATIONSHIPS

Having determined the segment content and the physical relationships between segments, the next task is to investigate the logical relationships that may be suggested by the Type M associations in the composite model. Specifically, candidates are determined for logical relationships and for secondary indexes.

The word *candidates* should be emphasized. These are merely places in the resulting logical model where logical relationships and secondary indexing should be considered. The final implementation decisions, which depend on the frequency of accesses to be made and on the volume of data to be searched, should be made by the designer. The contribution of the automated procedures is to alert the designer to places where such alternatives exist and to provide performance information that can help in making the decisions.

Determine Candidates for Logical Relationships

Logical relationship candidates are determined in three ways:

- M:M mappings between keys
- Type M association between keys
- Unrelated multiple parents of a segment

M:M Mappings Between Keys

When two keys are related by M:M mappings, a bidirectional relationship is defined between their segments and can be so reported. Further, the presence or absence of a third element to define the intersection can also be reported. Although their attributes are not shown, Figure 7-1 assumes that A, B, and C are keys.

It should be noted that an element defining an intersection need not have been specified in advance. The procedures can not only detect candidates for logical relationships, but can also inform the designer that the needed intersection is already present or that it still needs to be specified.

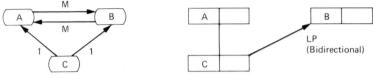

Figure 7-i.

Type M Association Between Keys

If the designer elects not to let the automated procedures supply a Type 1 inverse association when a lone Type M association exists between keys, he presumably wants such situations implemented, or at least considered, as logical relations. The automated procedures will treat such cases the same way as M:M mappings. For implementation, a unidirectional logical relation will suffice, although the bidirectional relationship may be preferred for extensibility reasons. Again, in Figure 7-2, A, B, and C are keys.

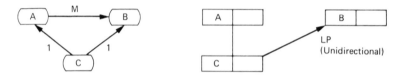

Figure 7-2. Logical relationship from Type M association.

Unrelated Multiple Parents of a Segment

In tracing the Type 1 associations to determine physical hierarchies, it is likely that some segments will have more than one possible parent even though no direct associations have been specified between these possible parents. This indicates that there is no requirement to traverse from one parent to the other. In these cases, logical relations may still be desirable in order to avoid redundant storage of the common child rather than to provide access paths between the parents. In Figure 7-3, A, B, and C are to be considered as keys.

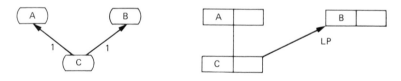

Figure 7-3. Logical relationship from multiple parents.

Selecting the Physical Parent

For illustrative purposes in the foregoing examples, one of the candidate parents was arbitrarily selected as the physical parent and the other as the logical parent. To a large extent, these choices can be made automatically by considering the relative importance, from a performance standpoint, of the paths to be traversed. (Performance weights, to be introduced in Chapter 10, can provide the measures of the relative importance of the paths.) However, if there are two or more logical parent candidates, the designer must resolve their implementation because DL/I permits only one logical parent.

For bidirectional candidates and unrelated multiple parents, the paths of highest relative importance can be used to specify which is to be the physical parent. In the second case above, that of a lone Type M association between the candidate parents, the "from" element should define the physical parent, and the "to" element should define the logical parent.

To illustrate the selection of logical relation candidates from the composite model, consider the model illustrated in Figure 7-4a—the example of Figure 6-12 with elements E and R and their associations added. This composite view suggests a requirement for a logical relationship as shown in Figure 7-4b.

With the M:M mapping between B and E, a bidirectional logical relationship is suggested. If there were only a single Type M association between B and E, a unidirectional relationship would be suggested. Both cases should be noted and reported along with the additional information of whether or not a third element (D in this case) is present to define the intersection. Finally, if there was no direct association specified between B and E, a segment (D) with two parents (B and E) would also suggest a logical relationship.

Determine Candidates for Secondary Indexes

Candidates for secondary indexes are determined in three ways:

- The root of a local view is not the root of the resulting hierarchy
- A Type M association is specified from an attribute to a key in the hierarchical path leading to the attribute
- An identity between two elements has been resolved by designating one of the elements as an attribute and, therefore, also as a candidate for secondary indexing

The Root of a Local View is Not the Root of the Resulting Hierarchy

Assume that User 1 submits a local view (Figure 7-5a) consisting of keys A, B, and C and that User 2 submits a local view (Figure 7-5b) containing keys

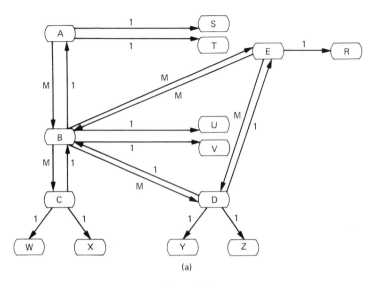

(a)

Figure 7-4a

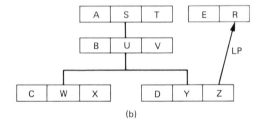

(b)

Figure 7-4b

Figure 7-4. Logical relationship from the composite model.

B, D, and E. The composite of these two local views and the resulting logical structure are shown in Figure 7-5c and 7-5d.

Since B was the root key of User 2's local view, but does not constitute the root segment of the resulting hierarchy, B becomes a candidate for secondary indexing.

A Type M Association from an Attribute to a Key

Consider the situation illustrated in Figure 7-6.

B is a key because it has a Type 1 association emanating from it, and A is an attribute of B. Thus, A becomes a nonsequence field in B's segment.

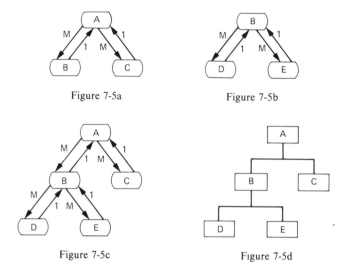

Figure 7-5a   Figure 7-5b

Figure 7-5c   Figure 7-5d

Figure 7-5. Secondary index from a nonroot key.

Figure 7-6. Secondary index from an attribute.

Because of the Type M association from A to B, A is listed as a candidate for secondary indexing.

This example is frequently misunderstood. Because of the M:1 mapping, one may tend to think of A and B as having a parent-child relationship. But when A is not identifying any other element with a Type 1 association, it is not considered to be a key, and therefore, represents neither a parent segment nor a segment.

Figure 7-7 shows that the assumed interpretation is reasonable and can be valid.

Figure 7-7a   Figure 7-7b

Figure 7-7. Interpreting 1:M mappings to an attribute.

In Figure 7-7a, PRICE is an attribute of PART-NO. It cannot be considered a key (parent of PART-NO) because there is no Type 1 association coming from it. The Type M association from PRICE to PART-NO indicates someone's requirement to find all parts of a certain price. The example in Figure 7-7b is analogous; someone wants a list of all players of a certain weight.

For DL/I implementation of secondary indexes, the key need not be in the same segment as the attribute, but it must be in the same hierarchical path and not at a lower level than the attribute.

An Element in an Identity

Two elements that identify each other uniquely (with Type 1 associations) are called an Identity. An example is EMPLOYEE-NO and SOC-SEC-NO (Figure 7-8). A common way to implement identities is to place both elements into

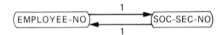

Figure 7-8. Secondary index from an identity.

the same segment with one element serving as the sequence field and the other element as the source of a secondary index. Identities are discussed more fully in Chapter 9.

8. Structuring Subtleties

In following the logical design procedures outlined in Chapters 6 and 7, certain subtleties can appear in the results which the designer should identify and diagnose. The intent of this chapter is to describe the subtleties that may be present and to indicate special care that should be exercised in designing the automated procedures. It will also indicate how the designer may recognize and interpret these situations and how he might resolve them. No claim is made that the problems discussed in this chapter and in Chapter 9 constitute a complete set of problems that can be encountered, but they represent all the problems thus far encountered by the author.

TYPE C ASSOCIATIONS

The Type C association is a special case of either the Type 1 or the Type M association, depending on the context in which it is used, and it will be treated as Type 1 or Type M by the automated procedures according to that context.

A Type C association to a nonkey element identifies an attribute value that may or may not be present in a segment. For purposes of computation, a Type C association, in this case, is processed as though it were a Type 1 so that the usual procedure for determining attributes and segment contents can apply (Figure 8-1). The target attribute can be flagged as conditional in the resulting segment.

A Type C association to a key indicates a target segment that may or may not exist. Such a segment cannot be a physical parent of the source segment because the physical parent must always exist. Therefore, the normal processing for a Type 1 association would be inappropriate in this case. A Type C association to a key is processed as though it were Type M and either of the two cases in Figures 8-2 and 8-3 is consistent with DL/I rules.

In Figure 8-2, A is structured into a parent segment with a child, B, that may or may not exist. In Figure 8-3, A and B are structured as parents in a candidate logical relation in which the B segment may or may not exist.

ROOT SEGMENTS WITHOUT ATTRIBUTES

It is not uncommon to have a data base with a root segment that has only a key and no attributes. In addition, large data bases sometimes have key-only segments at intermediate levels to serve as pseudoindexes into long twin chains

Figure 8-1. Interpreting a Type C association to an attribute.

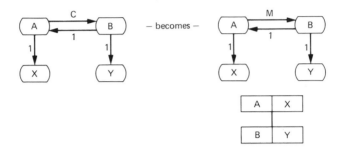

Figure 8-2. Interpreting a C:1 mapping between keys.

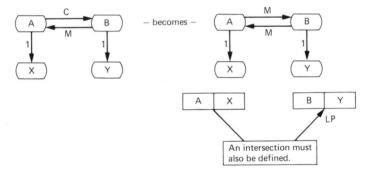

Figure 8-3. Interpreting a C:M mapping between keys.

to reduce search time. But the structuring procedures thus far outlined require an attribute before a root data element can be classed as a key.

As possible solutions, the designer may define a dummy attribute to satisfy the structuring algorithm's requirements, or the procedures may be designed to respond to a special structuring command to force a named element to be regarded as a key. This and other special structuring commands will be considered in Chapter 12.

REPEATING ATTRIBUTES

Floating elements as shown in Figure 8-4a may be a case of repeating attributes. A is a key and therefore a segment; B is not fully defined as a key or an

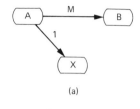

(a)

Figure 8-4a

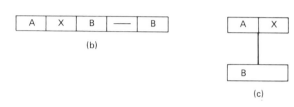

(b)

(c)

Figure 8-4b　　　　　　　　　　Figure 8-4c

Figure 8-4. Interpreting repeating attributes.

attribute because there are no Type 1 associations into or out of B. A is identifying several possible B values. B can be placed into A's segment as a repeating group field (Figure 8-4b), or it can be implemented as a dependent lower-level segment (Figure 8-4c). Either of these interpretations could be valid.

Element B can not be included in the resulting logical model without making an arbitrary choice. These situations should be reported on a diagnostic report and the designer should be asked to supply further defining associations or dictate the preferred alternative by means of a special structuring command.

M:M MAPPINGS

An M:M mapping between two keys, A and B, implies a logical relationship between A and B. But it is not always the case that A and B will be the actual physical and logical parents of that logical relationship.

If there is a third element, C, such that C is direct descendant of both A and B, then A and B are the physical/logical parent, with C serving as a direct intersection as in Figure 8-5.

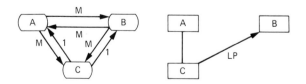

Figure 8-5. Interpreting M:M mappings with direct intersection.

Consider the example of Figure 8-6 with an M:M mapping between A and D. Here, F serves as an indirect intersection of A and D. These requirements will be structured as shown. In the resulting structure, C and E are the physical/logical parents of the logical relation with F as the intersection. But the structure supports the M:M mapping between A and D.

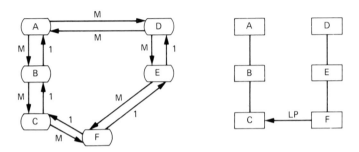

Figure 8-6. Interpreting M:M mappings with indirect intersection.

INVALID SECONDARY INDEX CANDIDATES

When determining candidates for secondary indexing according to a Type M association from an attribute to a key, the procedure, as described in Chapter 7, need not require that the key and attribute be in the same segment. While keys at an equal or higher hierarchical level in the same path as the source attribute are suitable targets for a secondary index, keys at a lower level or in a different path are a violation of DL/I rules for secondary indexing. Invalid key-attribute combinations, if reported at all, should be appropriately flagged.

INTERSECTIONS AT THE HIGHEST LEVEL

As discussed in Chapter 7, the intersection of a logical relationship is a common element from which paths of Type 1 associations originate and lead to each of the keys of the physical/logical parents. But, from an automated procedure, some superfluous intersections could be reported. For example, in Figure 8-7, C, D, and E are cascading intersections of A and B because each could be considered the origin of paths of Type 1 associations leading to A and B. The automated procedure should be able to detect the element where the paths diverge and list only that element (C in this case) as the intersection.

REPORTING LOOPS

A loop is a situation in which a path of Type 1 associations circles back upon itself. The simplest type of loop, the two-element identity, has previously been

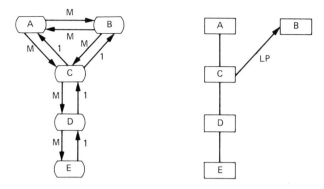

Figure 8-7. Interpreting M:M mappings with cascading intersections.

mentioned. Loops can represent valid situations within an enterprise (bill of materials, where-used, etc.), or they can represent incompatible views of the data which become apparent when several local views are merged together into a composite model. Therefore, all detected loops should be reported to the designers for their consideration.

If a loop is invalid, the designers must respecify a portion of the data requirements. If valid, he must specify where it is to be broken so that the correct parent-child relationships can be deduced. In this latter case a logical relation or redundant data is required to complete a loop in DL/I structures.

The subtlety in reporting loops is that several paths, each starting with a different element, can lead into the same loop as shown in Figure 8-8.

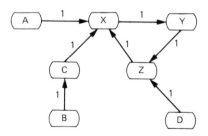

Figure 8-8. Example of a loop.

Unless appropriate checks are built into the automated procedure, the loop in this example could be reported seven times (Figure 8-9a). A single reporting of such a loop, as in Figure 8-9b, will be much less confusing to the designer.

X → Y → Z
Y → Z → X
Z → X → Y
A → X → Y → Z
C → X → Y → Z
B → C → X → Y → Z
D → Z → X → Y

Figure 8-9a

X → Y → Z

Figure 8-9b

Figure 8-9. Reporting loops.

LOGICAL RELATIONS BETWEEN THREE OR MORE PARENTS

Three logical relations are suggested in the bubble chart of Figure 8-10. The automated procedures, as outlined, will report X as an intersection between A and B, and Y as an intersection between B and C. But they will report the lack of an intersection between A and C. Yet it is possible to traverse from A to C in a DL/I implementation, provided the necessary occurrences of each element are available. The automated procedures should be designed to trace the M:M mappings sufficiently to note and properly report these types of logical paths.

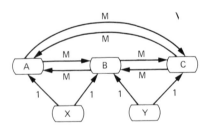

Figure 8-10. Logical relationships between three sets of parents.

9. Intelligent Structuring

THE CONCEPT

Chapter 8 addressed certain subtleties that can arise when following the automated procedures for logical design, and the discussion was oriented toward proper interpretation by the designer and toward proper design of the procedures to avoid reporting confusing diagnostics. This chapter is devoted to the identification and possible resolution of subtleties of a different nature. There are certain situations in which the structuring algorithms can produce results which are functionally correct, but which may be undesirable from an efficiency standpoint. Some of these situations can be identified automatically in context and, by making certain departures from the normal structuring algorithms, the procedures can derive a logical design that may be closer to what the designer will prefer for efficient processing. The trade-off is that the designer may be giving up a degree of data independence, especially with regard to possible future applications or changes. The designer must realize that in allowing some of these choices, made to promote processing or storage efficiency, segments may be produced containing some of the update anomalies that normalizing (removing transitivity) seeks to avoid. These anomalies are discussed in Date, Chapter 9 (Reference A.4).

This capability should be used with caution and, if used, the designer must be able to review the automated choices and override them in a subsequent iteration when necessary. The refinement techniques to be presented in Chapter 12 provide a means of making these overrides. This chapter reviews situations that are likely to occur and presents the rationale for the choices that an automated procedure might make.

Undesired Key-Only Segments

While key-only segments are sometimes desired in the resulting logical model, it is more likely that the designer has specified data requirements in such a way that the automated procedures have produced unwanted segments containing only keys. The section on deriving binary relations in Chapter 4 indicates how data requirements may be specified to reduce the occurrences of these types of situations.

There are two cases to consider: lone Type 1 associations and M:1 mappings.

Lone Type 1 Associations

If a lone Type 1 association exists from a key-only segment to a parent key (Figure 9-1a), the automated procedures will first attempt to place the data element of the key-only segment into the segment of its parent as an attribute (Figure 9-1b). If a clearly intended parent segment with at least two attributes is found, this is the action that will be taken.

Figure 9-1a   Figure 9-1b

Figure 9-1. Preferred removal of key-only segments.

The requirement of two or more attributes for the parent can be understood by referring to Figure 9-2a below. It is likely that a key with only one attribute was not really intended as a key, as is the case illustrated with STATE and COUNTRY.

Now, consider the local view of Figure 9-2a which is depicted with only Type 1 associations. This could be the result of a local view in which the designer does not intend attributes for any key other than EMPLOYEE. He simply wants to know the city, county, state, and country in which an employee lives. But the designer would draw the bubble chart of Figure 9-2a if he used the following rationale for specifying the binary relations: the employee lives in a certain city, the city identifies the county, the county identifies the state, and the state identifies the country.

By following the paths of Type 1 associations to determine segments and their parent-child relationships, the automated procedures, as outlined in Chapter 6, would derive the logical model depicted in Figure 9-2b. Remember that the Type 1 association between keys defines a relationship from child upward to parent. Remember also that COUNTRY, which has no Type 1 association emanating from it, will be considered an attribute rather than a parent of STATE. But because the designer has not indicated a need for other information about cities or about counties, etc., his processing needs can be satisfied naturally and more efficiently by considering these elements as attributes of EMPLOYEE and by structuring them into the EMPLOYEE segment, as shown in Figure 9-2c. By the associative rule of graph theory, each of these elements is uniquely identified by EMPLOYEE and can validly be compressed into the EMPLOYEE segment.

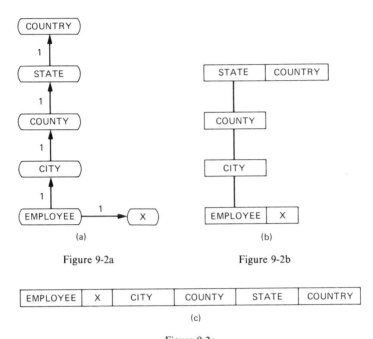

Figure 9-2a   Figure 9-2b

(c)

Figure 9-2c

Figure 9-2. Interpreting lone Type 1 associations.

EMPLOYEE is the only element for which other attributes (X) are specified; therefore, a single employee segment will suffice and will still preserve the specified relationships. Note that the specified Type 1 associations are not violated. An occurrence of the real source element (EMPLOYEE) still uniquely identifies an occurrence of each of the target elements.

The structuring rule is that in paths of lone Type 1 associations, with no inverse associations having been specified, each key element that has a single Type 1 association into it and a single Type 1 association emanating from it will be treated as an attribute in the segment of the closest predecessor key in that path that has attributes. For the example at hand, the result will be that shown in Figure 9-2c.

Key-only segments that are targets of two or more Type 1 associations from other keys should not be automatically eliminated but should be noted for the designer's review and evaluation. If the key-only segment is a common parent, unwanted redundancy could be created by removing the segment and placing the key element into each of the lower-level segments.

The designer should carefully review the diagnostic list of key-only segment resolutions to see if the automatic retention or removal of such segments meets his needs.

M:1 Mappings

For key-only segments in paths defined by M:1 mappings (Figure 9-3a), the application programs will traverse from A to B to C to D and in the inverse direction. Therefore, the keys should continue to be treated as keys rather than as attributes. But levels of the resulting hierarchy can be eliminated without loss of information and without creating additional redundancy by concatenating the key field of a key-only segment into the key field of the next lowest segment (Figure 9-3b). If the keys of the eliminated segments were concatenated upward into the A segment, additional redundancy could be created because additional replications of the A segment occurrences could be required.

Key-only segments at which paths converge from below, or from above, should not be automatically eliminated. And again, the designer must be able to review and override, if necessary, the structuring results. For example, key-only segments are sometimes desired as an indexing device for reducing the time to search long twin chains.

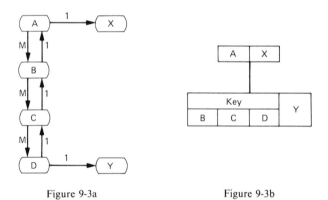

Figure 9-3a   Figure 9-3b

Figure 9-3. Interpreting M:1 mappings between key-only segments.

Lone Type 1 Association to a Key that has Attributes

Whenever a single Type 1 association to a key exists with no corresponding inverse association to specify a requirement to traverse the opposite direction, the segment represented by the target key could be desired as a child rather than as a parent of the segment of the source key. Such a child segment type will be considered to have only one occurrence. This type of structure could provide processing efficiency by removing infrequently used information from the source segment. However, the procedures as described will interpret a Type 1 association between keys in a child-to-parent sense, and will produce a structure as illustrated in Figure 9-4.

Figure 9-4. Interpretation of lone Type 1 association to key with attributes.

An additional degree of intelligence in the automated procedures could identify segments pointed to by a lone Type 1 association and could structure such segments as children of their source segments (Figure 9-5). On the other hand, the designer may prefer to eliminate the target segment by placing its key and attributes into the source segment. But either of these alternatives could lead to unwanted results if applied globally to all local views. This kind of intelligence, if used, should be optional globally or by named local views. The designer should request it or bypass it according to his understanding of his needs. Regardless of the option chosen, a diagnostic list of lone Type 1 associations to keys should be obtained and reviewed by the designer.

Figure 9-5. Alternate interpretation of lone Type 1 association to key with attributes.

Unnecessary Levels from Compound Keys

It is possible that compound keys, though correctly specified, will cause the creation of unnecessary levels in the resulting hierarchy. Suppose, for example, that a department is engaged in several projects, and the work location of a given employee depends also on the project on which he is working (Figure 9-6).

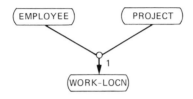

Figure 9-6. Example of compound keys.

Now suppose that each employee can be assigned to more than one project and that each project is staffed by several employees. The bubble chart of Figure 9-6 becomes that shown in Figure 9-7a, and the resulting structure, assuming EMPLOYEE and PROJECT both to be keys, will be as shown in Figure 9-7b.

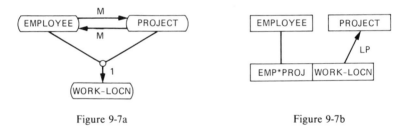

Figure 9-7a   Figure 9-7b

Figure 9-7. Interpreting compound keys related by an M:M mapping.

But suppose, instead, that each employee is assigned to only one project. Then the bubble chart of Figure 9-6 becomes that shown in Figure 9-8a, and the resulting structure will be as shown in Figure 9-8b.

In this second case, because EMPLOYEE and PROJECT are in the same hierarchical path, the compound key at the third level does not add any further qualification. Therefore, it can be eliminated, and WORK-LOCN can be placed into the EMPLOYEE segment.

The rule is that when simple elements of a compound key turn out to be in the same hierarchical path, a compound of those elements is not needed. But

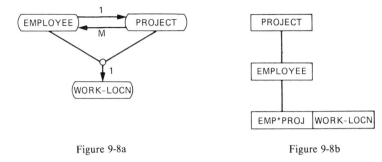

Figure 9-8a   Figure 9-8b

Figure 9-8. Interpreting compound keys related by an M:1 mapping.

when simple elements of a compound key appear in different hierarchical paths, the compound key is needed to represent the intersection. This type of intelligence can be built into the structuring programs, and occurrences of such situations and resolutions should also be reported to the designer for review.

The discussion of compound keys in Chapter 4 shows, in part, when compound keys should and should not be used in specifying the data requirements. But despite how a given local view is specified, the merging of different local views into a composite view can yield situations that may need to be refined in a subsequent iteration.

Resolve Identities

Two keys in a 1:1 mapping with each other constitute an identity as illustrated in Figure 9-9. Each key uniquely identifies the other (e.g., EMPLOYEE-NO and SOCIAL-SEC-NO). Since the Type 1 association between keys defines a child-parent relationship, an anomaly exists as to which key should be the parent and which should be the child.

Identities are generally implemented in one of three ways:

1. Both keys, and their attributes, may be placed into the same segment. One of the keys, usually the one used most frequently, is designated as the primary key and the other becomes the source of a secondary index.

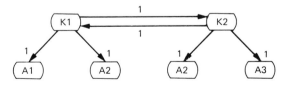

Figure 9-9. Example of an identity.

2. The two keys may be implemented as separate segments in a parent-child relationship. The key that is the source of the most frequently used association usually becomes the parent.
3. The two keys may be declared synonyms for the same data item, one of them simply eliminated, and the attributes of both placed into the same segment.

Experience indicates that identities are more frequently implemented in a single segment than in a parent-child relationship. Therefore, when an identity is detected, the automated procedures can attempt to resolve it by placing both keys and their attributes into the same segment (Figure 9-10).

K1	K2	A1	A2	A3

Figure 9-10. Usual resolution of an identity.

The key having the highest frequency of use as indicated by performance weights (see Chapter 10) can be selected as the primary key. Otherwise, if performance weights are not available, K1 will be selected as the primary key if:

- K2 has no attributes (Figure 9-11)
- K2's attributes are a subset of K1's attributes (Figure 9-12)
- K2 has fewer attributes than K1 (Figure 9-13)

Figure 9-11. Resolving identities (1).

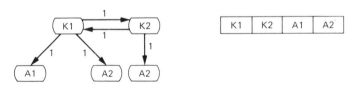

Figure 9-12. Resolving identities (2).

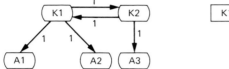

Figure 9-13. Resolving identities (3).

In all cases, the key not selected as the primary key will be listed as a candidate for secondary indexing. And again, all instances of identities should be reported to the designer for review and possible respecification in a subsequent iteration.

10. Performance Weights

If certain performance-oriented information can be entered into the design procedure with the binary relations of the local views, Performance Weights can be calculated that will provide a measure of the relative importance of the hierarchical paths in the derived logical model. These are dimensionless weights to be used in a comparative sense. For example, if the performance weight for Path A is ten times larger than the performance weight for Path B, then Path A can be considered to be ten times more important than Path B from a performance standpoint. Note that we are not saying, necessarily, that Path A is traversed ten times more often than Path B. The calculated number of traversals can be weighted by several factors as will be explained.

BASIC INPUTS FOR PERFORMANCE WEIGHTS

The basic inputs for performance weights calculations are the setup period and frequency of processing for each function (local view) and the expected frequency of use of the association paths to be traversed within each function.

Setup Period and Processing Frequency

How frequently and over what time periods are the functions expected to be invoked? A payroll may be processed twice a month. A personnel file may be updated once a week. Customer service inquiries may be performed 500 times a day on the average, or at a maximum rate of 150 times an hour. These processing frequencies and their setup periods provide the rates at which the various functions are expected to be invoked.

Frequency of Use of the Association Paths

For each function that is invoked, the following questions exemplify the frequency of use considerations for the paths to be traversed by the function. For each path to be traversed, how many accesses will be made to the primary (root) identifier? And for each of these accesses, how many accesses are expected to the next element on the path? And for each access to that element, how many accesses are expected to the element at the next level, and so on?

Example

An example of the foregoing concepts is some information requirements of a football league involving daily inquiry into a player's status and weekly updates of the players' records.

Consider first an inquiry into the positions a player can play and his ranking for each of those positions. The bubble chart of Figure 10-1 describes this local view.

Local View A

Figure 10-1. Local view A.

Assume each inquiry is for a single player and that an average of ten such inquiries are made hourly. The setup period is hourly and the processing frequency is ten. Assume further that the average player can play three positions and that for each position there is only one set of status (including rank) information. These individual association accessing frequencies can be indicated on the bubble chart as shown in Figure 10-2. The indication is that for each player, the association from PLAYER-NO to POSITION is traversed an average of $1 \times 3 = 3$ times, and the association from POSITION to RANK is traversed an average of $1 \times 3 \times 1 = 3$ times. These are the resultant frequencies of use of the associations.

Figure 10-2. Local view A with expected average accessing frequencies.

The basic input, then, for performance weight calculations for this local view is given in Figure 10-3.

From this basic input, the automated procedures will calculate the Resultant Frequency of Use (Figure 10-4) for each association for each time the function is invoked.

Setup Period:	Hourly
Processing Frequency:	10
Traversals (Frequency of Use):	
Root Key (PLAYER-NO)	1
PLAYER-NO → POSITION	3
POSITION → RANK	1

Figure 10-3. Performance weight parameters for local view A.

PLAYER-NO → POSITION	$1 \times 3 = 3$
POSITION → RANK	$1 \times 3 \times 1 = 3$

Figure 10-4. Resultant frequencies of use for local view A.

CALCULATING PERFORMANCE WEIGHTS FROM THE COMPOSITE MODEL

Background

Assume now that the football league also has a weekly batch function (once a week) of providing each team with a list of its player activity for the previous week. The bubble chart of Figure 10-5 describes this local view.

Local View B

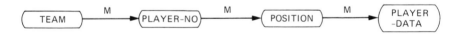

Figure 10-5. Local view B.

Assume that the league has 30 teams, and that the report is produced for all teams. Assume further that each team has 60 players, but only an average of 45 players play in a given game. Also assume that of the players who appear in a game, each plays 2 positions during the course of the game. Each player will appear on the report, but position and player data will be provided only for those players who played in the game. Thus we expect 2 traversals from PLAYER-NO to POSITION for 45 of the players and no traversals for the remaining 15 players. This means an average of $(45/60) \times 2 = 1.5$ traversals from PLAYER-NO to POSITION. And there are 5 types of player data per player. These frequencies may be indicated on local view B's bubble chart (Figure 10-6).

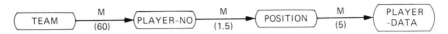

Figure 10-6. Local view B with expected average accessing frequencies.

The basic input for performance weight calculations for local view B is given in Figure 10-7.

Setup Period:	Weekly
Processing Frequency:	1
Traversals (Frequency of Use):	
Root Key (TEAM)	30
TEAM → PLAYER-NO	60
PLAYER-NO → POSITION	1.5
POSITION → PLAYER-DATA	5

Figure 10-7. Performance weight parameters for local view B.

The Resultant Frequency of Use of each of these associations, as calculated by the automated procedures, is shown in Figure 10-8.

TEAM → PLAYER-NO	30 × 60 = 1,800
PLAYER-NO → POSITION	30 × 60 × 1.5 = 2,700
POSITION → PLAYER-DATA	30 × 60 × 1.5 × 5 = 13,500

Figure 10-8. Resultant frequencies of use for local view B.

Combining these two local views gives the composite model of Figure 10-9. Note that the association from PLAYER-NO to POSITION was used in both local views. To get the total frequency of use for this association, we would like to add its resultant frequencies from each of the local views. But in one case the frequencies are on an hourly basis, and in the other case they are on a weekly basis. Therefore, the frequencies must be normalized to a common

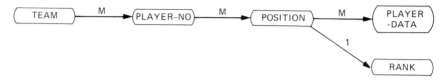

Figure 10-9. Composite model from local views A and B.

time period before they can be meaningfully added. Suppose "week" is chosen as the normalization time period.

Local view B is already in terms of a week. Local view A, which is in terms of hours, must be normalized to an equivalent weekly basis. Assuming that the league normally operates on the basis of an eight hour day and a seven day week, the time normalization is performed by multiplying all of local view A's frequencies by $8 \times 7 = 56$.

Other normalizations, or weightings, can also be performed. The entire calculation process is illustrated below.

Performance Weight Calculation

The means given here for calculating performance weights is suggestive rather than definitive, and can be done in any manner meaningful to the designer. The calculations are presented in a step-by-step manner, not because of complexity, but rather to assist the reader in judging how each step may best be applied in his own operating environment. Starting with the resultant frequencies of use for the associations in each local view, we have:

Local View A		Local View B	
PLAYER-NO → POSITION	3	TEAM→ PLAYER-NO	1,800
POSITION → RANK	3	PLAYER-NO → POSITION	2,700
		POSITION → PLAYER-DATA	13,500

The inquiries of local view A are processed on the average of ten times per hour. The function of local view B is invoked once a week. Therefore, the weights of local view A are multiplied by ten; the resulting weights are the total numbers of traversals of each association for their respective setup periods.

Local View A		Local View B	
PLAYER-NO → POSITION	30	TEAM → PLAYER-NO	1,800
POSITION → RANK	30	PLAYER-NO → POSITION	2,700
		POSITION → PLAYER-DATA	13,500

The weights of local view A are still in terms of hours, and the weights of local view B are in terms of weeks. We normalize local view A to a weekly basis by multiplying each of its weights by 56 (seven days at eight hours per day).

Local View A		Local View B	
PLAYER-NO → POSITION	1,680	TEAM → PLAYER-NO	1,800
POSITION → RANK	1,680	PLAYER-NO → POSITION	2,700
		POSITION → PLAYER-DATA	13,500

Assume that all accesses are retrieves except to PLAYER-DATA which will be updated one-half of the times it is accessed. In DL/I, this means that one-half of the accesses to PLAYER-DATA are GETs and the other half are GETs followed by REPLACEs. The weight for the association from POSITION to PLAYER-DATA is multiplied by 1.5 to account for this processing option.

Local View A		Local View B	
PLAYER-NO → POSITION	1,680	TEAM → PLAYER-NO	1,800
POSITION → RANK	1,680	PLAYER-NO → POSITION	2,700
		POSITION → PLAYER-DATA	20,250

Finally, for this example, online work is usually treated at a higher priority than batch work, and the designers choose to weight local view A by an additional factor of 5.

Local View A		Local View B	
PLAYER-NO → POSITION	8,400	TEAM → PLAYER-NO	1,800
POSITION → RANK	8,400	PLAYER-NO → POSITION	2,700
		POSITION → PLAYER-DATA	20,250

Now the performance weights for occurrences of the same association may be added. The resulting set of performance weights for this example are as follows:

TEAM → PLAYER-NO	1,800
PLAYER-NO → POSITION	11,100
POSITION → PLAYER-DATA	20,250
POSITION → RANK	8,400

ON SELECTING FREQUENCIES OF USE

Types of Estimates

Several approaches to estimating frequency of use of the association occurrences should be considered. The frequencies can be chosen according to maximum number of accesses, average number of accesses, average accesses over the peak period, or even by a statistical distribution. The choice depends on how critical performance is to the system being implemented and on how accurately the estimates can be made. Experience indicates that designers tend to work with maximum values of occurrence to determine space requirements and with average values of accesses to determine performance capability.

Selective Estimates

As a final thought, it is often difficult to obtain meaningful accessing estimates for all associations early in a design study. A more practical approach may be to do some initial design iterations, identify the critical areas that might be influenced by performance weights, and then make only those estimates necessary to support those areas. Considerable effort can be saved in this manner and the accuracy of the estimates is likely to be better.

11. Design and Diagnostic Reports

Two classes of reports should be provided by the automated logical design procedures. A series of design reports will present the results of logical structuring, and a series of diagnostic reports can provide helpful information and insight about the design. Remembering that the procedures are used in an iterative process, these reports should not be interpreted initially as documenting *the* logical design. After producing an initial logical model (following the editing iterations), additional iterations will probably be used to converge to the design to be implemented. As has been previously mentioned, the initial logical design will be functionally correct. But because of the subtleties that may be present, and because of design alternatives that should be resolved by the designer, iterative refinements will probably be made in which the designer specifies structural changes to be made. In Chapter 12 a full treatment of the "refinement" concept is given.

DESIGN REPORTS

The Design Reports document the logical model as derived by the automated procedures. They can also be useful as documentation of the process of converging to this final design. Throughout the entire process, two of the design reports are intended to be suggestive rather than definitive. The Candidates for Logical Relations and the Candidates for Secondary Indexing reports show where logical relations and secondary indexing are suggested by the design criteria. The choice of which ones to implement belongs to the designer. The reports merely show all places that should be considered, and provide supplementary information, such as performance weights, to help the designer make the choices.

The following four design reports are recommended:

- Parent-Child graph
- Suggested Segments
- Candidates for Logical Relationships
- Candidates for Secondary Indexing

The purpose and content of these reports are suggested without attempting to suggest their formats.

Parent-Child Graph

The derived logical structure of segments and the physical hierarchies into which they are organized can be presented graphically as in Figure 11-1. Indications of logical relations and secondary indexes could also be shown. The numbers in parentheses are the performance weights for each leg of the hierarchy where they are available. For the first iteration, at least, the names inside the segments will be the names of the keys, because the segments cannot be named by the designer until after they have been determined. In the refining iterations, the designer should be able to enter segment names and have them appear on this report, either with or instead of the key names.

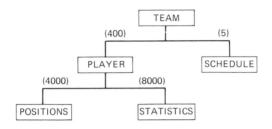

Figure 11-1. Parent-Child graph.

Suggested Segments

This is a listing of keys and the attributes clustered about each key. Each key and its attributes constitutes a suggested segment in the canonical logical model. Segment length can be presented as the sum of the lengths of the individual fields. If the expected number of occurrences of each element is included in the data requirements, then the expected number of segment occurrences can be calculated. From this information, the total number of data bytes for each segment type and the total number of data bytes for the data base as a whole can also be included in the report.

Redundant attributes should be flagged and performance weights from key to attribute should be presented. Also, appropriate notation can be made for conditional attributes, variable length attributes, repeating attributes or grouped attributes. Other attribute characteristics such as length, type, format, and processing option can also be shown.

For example, if suggested segments were to be derived from the binary relations of Figure 11-2a, the two segments of Figure 11-2b would be obtained. The report should clearly indicate that Y is a redundant attribute appearing in more than one segment, and that X is an attribute whose value may or may

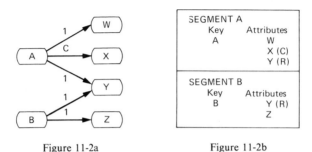

Figure 11-2a   Figure 11-2b

Figure 11-2. Reporting suggested segments.

not exist for each occurrence of its key, A. The other field and association char-acteristics can also be appropriately displayed.

Candidates for Logical Relations

This is a list of segments between which logical relations should be considered. Three types of situations can be reported. One will be a list of keys in an M:M mapping (Figure 11-3a). Another will be a list of keys related by a lone Type M association (Figure 11-3b). To successfully implement either of these two cases as parents in a DL/I logical relationship, a third element type represent-ing an intersection must also be specified to define paths from a given occur-rence of A to a given occurrence of B, and vice versa. This third element will usually be implemented as a logical child of the logical (destination) parent, although it could be defined in such a way as to be implemented as a common parent (rather than as a child) of both A and B. The presence or absence of such an element should be reported, and its role, if it is present, can also be reported. The third situation to be reported is that of a key having multiple parents without an M:M mapping or a Type M association between its parents (Figure 11-3c).

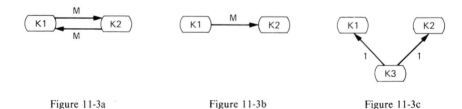

Figure 11-3a   Figure 11-3b   Figure 11-3c

Figure 11-3. Sources of candidates for logical relations.

Candidates for Secondary Indexing

The fourth recommended design report is a list of candidates for secondary indexing. As was explained in Chapter 7, a data element is selected for this report according to either of three criteria.

1. Root keys in local views that are not roots of the resulting hierarchies.
2. Attributes that refer to keys, or to other attributes, at an equal or higher hierarchical level with a Type M association. Situations in which a target element is at a lower level or in a different hierarchical path than its source element could also be reported and appropriately flagged.
3. Elements designated as secondary indexing candidates when resolving identities.

DIAGNOSTIC REPORTS

The wealth of diagnostic information that can be produced may be of even greater value to the designer than the design reports. Errors, inconsistencies, omissions, and design alternatives can be reported. Such information, available early in the design study, can help the designer complete the design in significantly less time and with greater likelihood of not overlooking situations that may become troublesome when the data base is implemented. The following is a representative list of the types of diagnostics, beyond the editing reports, that can be obtained. All of these reports, with two exceptions—Data Element Usage Matrix and Structure Specification Command Log—to be discussed in Chapter 12, have already been explained in Chapters 6–10. Therefore, the following list will carry minimal descriptions.

Nonessential (Transitive) Associations

A list of the transitive associations identified and removed from the design. The designer may want to reinsert some of these associations, at the expense of redundancy and update anomalies, to gain performance and validity benefits or both. The designer may also resolve them by considering some of the elements as homonyms and by making appropriate naming changes.

Identities

A list of data elements in a 1:1 mapping. These elements, as keys, cannot be mapped into a hierarchical structure without further definition.

Path Rule Violations

A list of data elements in paths of Type1 associations that loop back onto themselves. Also a list of paths having more than 15 elements. Both these situations violate DL/I rules for paths in a physical hierarchy.

Repeating Attributes

A list of floating elements that could be structured as repeating fields in a segment or as a dependent nonkeyed segment. These, generally, are nonkey elements identified by only a Type M association. They could also be structured into a dependent segment. The choice should be made by the designer.

Key-Only Segment Resolutions

A list of elements that would have been mapped into key-only segments because each has a Type 1 association to another key, but does not have a Type 1 association to an attribute of its own. The elements have been structured, instead, as attributes of their closest identifying keys.

Lone Type 1 Associations to Keys

A list of Type 1 associations to keys where no inverse association is specified. These define child-to-parent relationships, but sometimes, unless there is a Type M inverse association, the designer would prefer to structure the target key as an attribute in the segment of the source key or as a child of the source segment.

Intersecting Attributes

A list of attributes having more than one key. The attributes and their keys may represent redundant or intersection data. They may also be the result of naming inconsistencies. This report was suggested previously as an editing report, but it is also suggested here because of the possible structural implications of its information.

Modified and Augmented Mappings

A list of Type M associations for which Type 1 inverses were automatically supplied. Also a list of Type C associations and whether they were treated as Type 1 or as Type M. The designer should review this list to judge whether or not the assumed associations are in harmony with the functional and intrinsic data requirements.

Local View Structuring

A report indicating the contributions of each refined local view in the logical design results as depicted in the design reports. This report would be valuable to the designer who wants to evaluate the effect of an application function(s) on the overall design.

Performance Weight Details

A presentation of performance weight calculations showing the contributions of each association occurrence. This report will be helpful in judging the performance implications of associations from a particular local view.

Data Element Usage Matrix

A matrix presentation showing which data elements are used by which functions. The presentation can be made by suggested segment or on an overall basis. This is essentially a condensed form of the where-used list from the editing phase and is tailored to help evaluate the appropriateness of the segment contents.

Structure Specification Command Log

A list of the structure specification commands used and the associations that were automatically added or deleted in support of each command.

12. Refining the Design

Data base design is an iterative process and the automated techniques described herein are intended to be used in an iterative manner. The initial logical model produced after all diagnostics are resolved is said to be in canonical form. For various reasons, the designer may want to make alterations to this canonical model to obtain results beyond the scope of the automated procedures.

The refining phase should not be misused by injecting design constraints that are incompatible with the needs of the application functions. Rather, it should refine the canonical design to better serve the needs of the applications. It is the designer's responsibility to see that the refined design still supports the true data requirements of each application function.

THE REFINEMENT CONCEPT

When the editing phase of conceptual design is completed, the design proceures move into the structuring phase of logical design, and iterate (under the designer's control) between structuring and reediting until all situations reported by the diagnostics are resolved. At this point, a canonical model has been derived. Refining this canonical model is also an iterative process, but of a somewhat different nature as indicated in Figure 12-1. The canonical model was derived primarily from the functional data requirements, and while it will provide the needed access paths, certain rearrangements of its structure may provide improved processing efficiencies or security features or other benefits to be discussed below. Thus the designer now moves into the mode of imposing structural constraints or rearrangements of the canonical logical model already obtained, and the automated procedures are now serving more for quality control and documentation than for design. The objective, now, is to identify and resolve any additional incompatibilities or alternatives arising from the new constraints or rearrangements being imposed. Instant documentation of the results is another advantage of using the automated procedures for processing these refinements.

The refinement concept emphasizes the true role of the logical design procedures. The procedures do not perform a logical design *per se*. Rather, they serve as a computational tool which organizes the data requirements into a canonical logical model and which performs documentation and quality control checks of the designer's refinements. In the refinement phase, the human

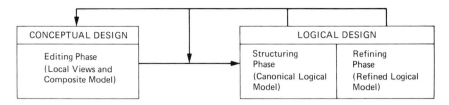

Figure 12-1. The refinement concept.

designer does the real logical design with the assistance of such a tool. The overall process is depicted in Figure 12-1.

Not all the refinements need to be reedited. As will be seen, some refinements may simply be structural constraints without adding to or changing the data requirements as specified. Therefore, the automated procedures should be desgned to permit refinement iterations to be invoked at the beginning of the structural phase and at the beginning of the editing phase.

REASONS FOR REFINEMENT

The designer may want to alter the canonical logical model for various reasons. Some of the more common reasons are discussed below. Techniques for making these alterations will be suggested later in this chapter.

Performance Benefits

Although the canonical logical model may fulfill the functional aspects of the data requirements, it may not be efficient from a performance standpoint. For example, suppose the structure of Figure 12-2a has been produced and that C is a candidate for secondary indexing because C is the root key of some of the required functions but not the root segment of the resulting structure. If most

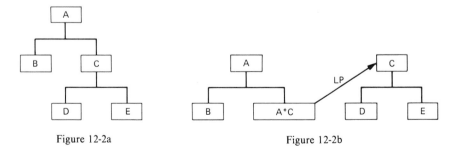

Figure 12-2a   Figure 12-2b

Figure 12-2. Example of a performance refinement.

entries into the data base are through its root segment, A, the design may be efficient. But if most entries are through C, then a performance improvement may be gained by making C and its children a separate data base and logically relating A and C (Figure 12-2b).

Another refinement for performance reasons is the addition of some redundancy to the data base in order to reduce the number of required accesses. For example, by placing a replicated value of a data element into a second segment, it may be possible to limit the accesses to one segment rather than to two or more.

Segment Rearrangement

The automated procedures place a key and all its attributes into a single "suggested" segment. The designer may want to rearrange some of these segments for various reasons. For example, the designer may wish to divide a suggested segment in such a way that the fields in one segment are used by one program and the fields in the other segment are used by a different program.

Segment types whose fields are subject to a high degree of insert or delete activity may be divided so that required information will not be unnecessarily deleted and so that unnecessary information will not be required for insertions.

Smith and Smith (Reference C.13), with their Principle of Aggregation, suggest that segments be subdivided because the relationship from the key to some attributes is semantically different from the relationship from the same key to other attributes. For example, if a certain student is enrolled in a certain class and has climbed a certain mountain, CLASS and MOUNTAIN could both be included in the STUDENT segment. But the relationships from STUDENT to CLASS and from STUDENT to MOUNTAIN represent different concepts, and a cleaner design might be obtained by splitting such a segment.

There are many other possible reasons for subdividing the suggested segments. The segment may be too long for reasonable buffering or work space allocation. Or the designer may want to move some fields into a separate segment for security purposes, to separate fixed length fields from variable length fields, or to separate frequently used fields from infrequently used fields.

On the other hand, the designer may want to combine some segments, at the risk of creating some redundancy, to reduce the number of levels in the hierarchy or to reduce the number of accesses at a given level.

Finally, the designer may choose to create new segments not directly suggested by the data requirements. For example, one commonly used technique for improving performance where long twin chains are involved is to create a new segment type as an immediate parent to be used as a pseudoindex into the twin chain.

New Design Insights

After analyzing the logical model and studying the interaction of several sets of data requirements, the designers or end users may obtain new perspectives and additional insights for considering the design. For example, paths may be present between data elements not directly related in any single local view. Or data elements considered by some users to be properties of entities may turn out to be entities themselves with their own properties.

Structural results that are different from, or in addition to, what the designer expected suggest new insights which sould be evaluated. Perhaps some outputs or functions can be combined or eliminated. Other efficiencies may be suggested by the design alernatives that are reported. An understanding of the paths and associations resulting after the local views are combined can help suggest new functions or queries that can be accomodated.

Overriding Decisions of the Automated Procedures

Although it is desirable that the automated procedures make certain choices of their own when possible, the designer must have the ability to review and override any of these choices. Nonessential associations removed from the design, the choice of physical parentage, and the "best guesses" outlined in Chapter 9 are examples of decisions the designer must be able to override.

Design Decisions by Intuition or by Edict

Sometimes certain design constraints are imposed simply because the designer "feels" they are right or because company policy demands it. For example, in many financial data bases, it is "company policy" that the root segment will be the account number.

Changing Requirements

Before most design studies are completed, requirements change. Some functions are no longer needed. New functions are wanted or, more commonly, existing functions are to be expanded and perhaps processed differently. The end user seems always to want to know what it will cost to add this or that capability. Although the degree to which changes can be tolerated depends on the individual situation, they nevertheless do occur, and the automated techniques are particularly suited for processing and evaluating such requests.

Data Base Management System Constraints

Unless proper constraints are built into the structuring procedures, it is possible to obtain a logical design which, while functionally correct for the data requirements, violates some rule of the data base management system to be used. For example, in designing for DL/I structures, it is possible that Candidates for Logical Relations will call for an implementation leading to more than one logical parent, or to a logical child of a logical child, both of which are violations of IMS rules. Suitable refinements are then necessary.

REFINEMENT TECHNIQUES

The following is a list of the more likely refinements to be desired in a design study, along with an indication of how they might be implemented. The automated procedures can be designed with a set of structure specification commands to enable the designer to request any of these refinements.

Change a Parent-Child Relationship to a Logical Relationship

Two segments in a physical parent-child relationship are to be made parents in a logical relationship, with an intersection segment defining the access path between them (Figure 12-3). The designer should be able to indicate, via the structure specification command, which of B's attributes stay with B and which should become intersection data.

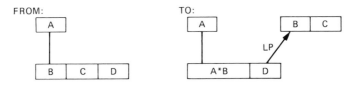

Figure 12-3. Parent-Child relationship to logical relationship.

Change a Logical Relationship to a Parent-Child Relationship

Two segments that are parents in a logical relationship are to be placed into a physical parent-child relationship (Figure 12-4).

Divide a Segment

A segment is to be divided with some of its attributes remaining in the origianl segment and other attributes going into the new segment. The new segment

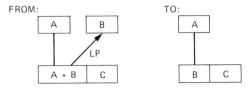

Figure 12-4. Logical relationship to parent-child relationship.

can either be at the same level as the one that spawned it or it can be a dependent segment of the original (Figure 12-5).

The structure specification command must specify which attributes will remain in the original segment, which will go into the new segment, and whether the new segment is to be a peer or a dependent of the original.

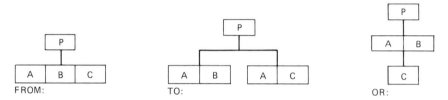

Figure 12-5. Divide a segment.

Combine Segments

Two segments in the same hierarchical path are to be combined into one (Figure 12-6). This means the possibility of redundant storage of the data from the lower-level segment. But it can also mean a reduction of the number of accesses to reach that data.

Figure 12-6. Combine segments.

Insert a Segment

A segment, possibly an indexing segment, is to be inserted into a path between a parent segment and some of its children (Figure 12-7).

Figure 12-7. Insert a segment.

Change Redundant Data to Intersection Data

Move data elements that appear redundantly in both parents of a logical relationship into the intersection segment (Figure 12-8).

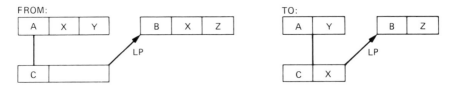

Figure 12-8. Redundant data to intersection data.

Select a Physical Parent

For a segment that has more than one possible parent, select a physical parent other than the one selected by the automated procedures (Figure 12-9).

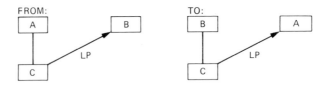

Figure 12-9. Select a physical parent.

Reinsert a Nonessential Association

Reinserting a nonessential association may mean reintroducing some of the update anomalies that normalizing seeks to avoid. But when performed in a controlled manner, such reinsertions may safely yield performance benefits. See Date, Chapter 9 (Reference A.4), for a discussion of these possible update anomalies.

From Key to Attribute

This means creating redundant data by adding an attribute to a segment that is at a lower level than the segment where the attribute is currently placed (Figure 12-10). This is sometimes desired when accessing B's segment via a secondary index or a logical relationship without going through A.

Figure 12-10. Reinserting a nonessential association (key-to-attribute).

From Segment to Segment

A segment is to be accessed directly from another segment that is at a higher level than the target segment's immediate parent, or a direct access is desired from the lower segment to the higher (Figure 12-11). One approach avoids a logical relationship by redundantly storing an additional copy of the C segment. The structure specification command should permit renaming the key. This alternative may be desired in retrieval situations where space is not a critical factor. A second approach, implemented with a logical relationship, avoids redundant storage of the segment and thus is better for update situations.

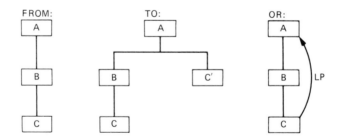

Figure 12-11. Reinserting a nonessential association (segment to segment).

Resolve a Repeating Attribute

Cause a floating (repeating) attribute to be placed as a repeating field into the segment of its key, or to be structured into a lower-level segment (Figure 12-12). The structure specification command must be able to specify the desired alternative.

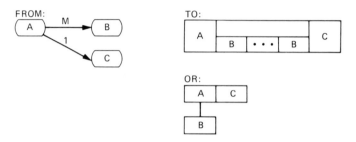

Figure 12-12. Resolve a repeating attribute.

Define an Intersection

Define an intersection for two parents of a candidate logical relationship (Figure 12-13). This situation arises when two keys are related by an M:M mapping, or by a single Type M association, and no third element has been specified with a path of Type 1 associations leading to each of the keys. The structure specification command must be able to designate the physical parent and any desired intersection data.

Figure 12-13. Define an intersection.

Resolve an Identity

The designer can choose to resolve identities (Figure 12-14a) by (1) placing both keys and all their attributes into one segment (Figure 12-14b), (2) placing the keys into a specified parent-child relationship (Figure 12-14c), or (3) eliminating one of the keys and placing its attributes into the other key's segment (Figure 12-14d). In the first alternative, one key becomes the primary key and the other becomes a candidate for secondary indexing. In the second alternative, the subservient key becomes a candidate for secondary indexing. The structure specification command can provide for selection of either of these three alternatives.

Force a Data Element to Be a Root Key

The data element, A, is desired as a root key. It may have been intended as a root key without attributes. If so, the automated procedures will have placed

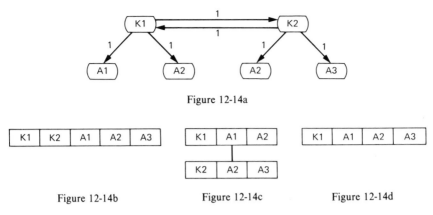

Figure 12-14a

Figure 12-14b   Figure 12-14c   Figure 12-14d

Figure 12-14. Resolve an identity.

it as an attribute into the segment of its intended child (Figure 12-15a). On the other hand, element A may have been structured as a valid key but not as a root key (Figure 12-15b). A structure specification command is needed to create a root segment for A, and if necessary, to logically relate the two resulting hierarchies.

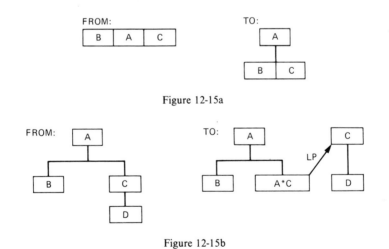

Figure 12-15a

Figure 12-15b

Figure 12-15. Force a data element to be a root key.

CRITERIA FOR GOOD DATA BASE DESIGN

In order to judge the appropriateness of the canonical logical model and the extent to which it should be refined, the designer needs some kind of established criteria against which to compare the model. The subjective benefit of experience, along with several judgment criteria are discussed below.

Experience

The importance of experience in judging the "goodness" of the logical model can hardly be overemphasized. Suppose, for example, the logical model contains the structure shown in Figure 12-16.

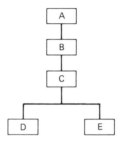

Figure 12-16. Possible bad design.

The designer's experience will tell him that this design may not give good performance if most of the accesses are to the lower-level segments because of the number of hierarchical levels, and he will look to see of some of the levels can be eliminated. To do this he may want to examine the performance weights for the path, the amount of redundancy that will be created, the storage space available, the amount of update activity, and so forth. These considerations can be made objectively, mostly from information provided by the automated procedures, but it was experience that caused the designer to suspect the design in the first place.

Experience alerts the designer to many possible design problems. For example, redundant data in the update mode can cause integrity and performance problems. Integrity problems arise in updating because, in actual practice, it has been difficult to keep all copies of the same data at a consistent level of currency. One kind of possible performance problem is that a large number of occurrences of a given segment type can cause lengthy twin chain searches. Certain logical designs can be inappropriate for insertion and deletion activity. The objective information provided by the automated procedures can assist the designer in these considerations, but it is not a substitute for experience.

Predesign Review

Many practitioners of data base design claim that before launching into a full design study, a cursory review of the functional requirements can provide a helpful framework for judging the adequacy of the resulting design. Sometimes called a Predesign Review, the exercise should not take longer than a few weeks, and should not descend to the data element level. It is simply a high-level overview of the functional data requirements.

To illustrate, consider an insurance company that plans to automate its contracts administration application in a data base environment. Contracts administration usually means executing functions for new business, policy endorsements, policy billing, etc. Starting with new business, the major entities can be identified without extensive effort. New business will deal with a description of the policy itself, with the insured, and with the agent making the sale. There will be personal data, medical data, and historical data about the insured (if the insured has or has had other policies with the company). So the designer begins constructing a gross logical design (Figure 12-17) from an intuitive standpoint.

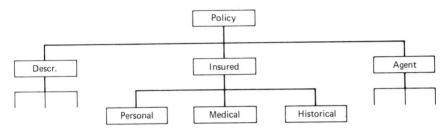

Figure 12-17. Intuitive structuring, predesign review.

In order to fully understand the functions and their requirements, the designer must do more than merely identify the major entities. He should also study the goals and the systems and procedures of the application area. He should become familiar with the work flow, the information flow, and their rules and constraints. For example, it may be important that, for all policies purchased after the 25th of the month, the billing function is not invoked until two months hence.

After considering the other functions of contract administration, the design may take on some logical relationships (Figure 12-18).

Now, having done this, the designer can launch into the fully detailed analysis of functional and intrinsic data requirements and, by following the systematic procedures that have been presented, he can derive a canonical logical design. Those parts of the canonical model that agree with his cursory intuitive approach can be accepted with confidence. But those parts that display a different structure should be carefully reviewed. The differences may be the result of functional requirements not analyzed thoroughly enough or they may reveal a structure superior to that which was intuitively conceived. The important

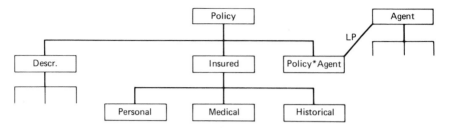

Figure 12-18. Alternate intuitive structuring, predesign review.

point is that the designer's attention is focused on those areas where additional consideration and review are in order.

The predesign review involves a series of interviews between the designers and end users. More specifically, it is recommended that the designers hold initial interviews with the managers of the application areas to obtain a gross overview of their requirements and ways of doing business. Then interviews with the application analysis can be more meaningful and efficient. Again, the entire process should be relatively brief. The purpose is to obtain an understanding of the user's working environment and to develop a general idea of a data base structure that will satisfy his processing requirements.

A major expected result of the design review is that the designer will significantly assist the end user in specifying his data requirements more clearly, more completely, and more consistently. But other, and perhaps even more important, results can come of these initial interviews—the rapport, trust, and confidence that can develop between the end users and the data processing personnel. The importance of these vital ingredients to the success of any data base design study cannot be overemphasized.

Data Element Usage Matrix and Its Use

With regard to segment content, the performance weights and processing options can be very helpful in determining if fields are assigned to segments in a manner consistent with good performance. The Suggested Segments can provide data element security codes, field lengths, and an indication of variable length fields. With this information, the designer can make additional judgments about the appropriateness of the suggested segment contents.

A Data Element Usage Matrix (Figure 12-19) is also a valuable tool for helping evaluate segment content. This is a where-used type of presentation showing which data elements (D.E.) are used with which application functions (A.F.). Statistical groupings can be obtained showing which elements are most frequently used together and by which applications and, on this basis, the sug-

EQUATION FOR PERFORMANCE WEIGHTS

For the above procedure, the equation for the performance weight of an association occurrence is:

$$P.W. = R.F.U. \times S \times T \times P \times O$$

where

    $R.F.U.$ = The Resultant Frequency of Use of the Association Occurrence
         S = Setup Period Frequency
         T = Time Normalization Factor
         P = Processing Option Factor
         O = Online Factor

SECONDARY INPUT FOR PERFORMANCE WEIGHTS

In addition to the basic input for performance weight calculations, some additional parameters are required as secondary input. The time periods of interest to the enterprise and the equivalences between time periods (e.g., 1 day = 8 hours, 1 month = 4.3 weeks) must be so specified. The normalization period must be selected. The equivalences between processing options (e.g., 1 insert = 10 gets) and the online factor must also be supplied. These factors are grouped together as secondary input because once selected, they normally will not vary for a design study. They can be retained by the system in a table for use with all iterations of a design study.

USING PERFORMANCE WEIGHTS

When calculated, performance weights can be used by the automated structuring for certain performance-oriented decisions. Due caution should be exercised by the designer, however, for these weights are only as good as the estimates from which they were calculated. Relatively safe decisions are the left-to-right ordering of children of a given parent, the choice of physical parent from a selection of alternate parents, and the relative use of attributes within a segment. Performance weights can also be helpful in more sensitive decisions to be made by the designer such as where to implement logical relations or secondary indexes, whether or not to combine or subdivide segments, and so forth.

	A.F.1	A.F.2	A.F.3	A.F.4
D.E.1	*	*		*
D.E.2	*		*	*
D.E.3		*		*
D.E.4	*	*		*
D.E.5			*	*

Figure 12-19. Data element usage matrix.

gested segment contents can also be evaluated. The matrix can be produced by segment or on an overall basis. Both presentations are useful.

The designer should be aware that a data element usage matrix does not normally reflect the functional relationships between the elements. It is performance-oriented but does not reflect inconsistencies in semantics and relationships, the anomolies that the normalization process seeks to avoid are likely to be present. Therefore, usage patterns should not be the sole criteria for composing segments.

Basic Checklist of Considerations

In judging the "goodness" of a data base design, the factors that generally receive the most attention are function, performance, and integrity. If the required data is present (or calculatable) and if it can indeed be accessed according to the rules of the data base management system to be used, the functional criteria is met. Performance is primarily concerned with how rapidly the required data can be provided to the application programs and, to a slightly lesser extent, with minimizing the frequency and impact of reorganization. Integrity is the concern about the accuracy and consistency of the data.

In addition to these more emphasized design criteria, a number of other factors should also be considered. These factors are primarily subjective in nature, and again, experience and intuition play an important arole in such judgments. The following considerations are representative of the evaluation concerns that are directly affected by the systematic procedures herein described.

1. Function
2. Performance
3. Integrity
4. Extensibility

5. Size
6. Security

Function

The local views define the elements and paths required for processing each function. These requirements are either built into the logical model derived by the systematic procedures or diagnostics are produced showing those that cannot be implemented. By resolving each of these diagnostics and using the procedures in an iterative manner to see if new diagnostics are created, the designer has a controlled mechanism for assuring a logical model that supports each of the processing functions.

Performance

Performance weights and frequencies of occurrence provide a means of making many performance-oriented judgments in producing and refining the canonical logical model. Based on these weights, choices of physical and logical parents and the left-to-right ordering of segment types can be made automatically. The performance weights can also help reveal such other performance-related concerns as the relative frequencies of use of fields assigned to the same segment or the relative frequency of use of a secondary index as opposed to direct entry through the root segment. These and other performance-oriented decisions could be built into the procedures, but it is recommended that, until considerable experience is gained in using these procedures, such considerations remain more a matter of human judgment; the procedures can provide helpful information for making those judgments.

Frequency of occurrence values for the data elements can also assist in performance considerations by providing a criteria for judging the lengths of twin chains.

While performance is the primary issue in physical design, these systematic procedures can assist in injecting good performance criteria into the logical design, prior to physical design processing. By making certain decisions automatically and by producing a nonredundant, canonical, logical model which the designer can then refine in a controlled manner, and with helpful diagnostics, these procedures can assist the designer in evaluating and controlling the performance implications of the candidate logical design models.

Integrity

Major assistance in assuring integrity has been provided by the conceptual design procedures of editing and producing lists helpful in identifying synonyms and homonyms and other problems in the meaning of the data.

Additional integrity assistance is available in the logical design procedures. Redundant data is identified on several of the lists that can be produced, and update intentions can also be identified. Thus the designer can be informed of all areas where update integrity may be a problem.

Integrity, in the sense of protection against loss or damage, is not directly addressed by these procedures.

Extensibility

Extensibility is the property of being able to support future requirements with the current design as a base. By describing the future requirements (as currently envisioned) in a separate local view, extensiblity is promoted. The "futures" local view shoold be worked into the design at a well chosen time in the iterative process, so the designers can evaluate any conflicts between current and future requirements. These systematic procedures provide a means of recognizing and resolving conflicts in a controlled manner.

Size

If the field sizes are included in the data requirements, sizes of the resulting data base and its segments can be derived. This information can be used by the designer in the refining process to determine whether to make some segments fixed or variable length or whether to reduce the number of hierarchical levels by combining a parent and child segment type. Other size considerations receive major emphasis in the physical design.

Security

By assigning security codes to data elements and to local views, design reports can be produced showing security sensitivity of data bases, segments, and fields. The procedures could be designed to make security-oriented decisions (e.g., segment content), but this is probably best done by the designer according to the needs of his organization.

13. Generating Relational Designs

Up to this point, the concepts of computer-assisted data base design have been presented in the context of designing DL/I hierarchical structures as implemented in IMS and in its Disk Operating System equivalent, DOS-DL/I. We will now lightly explore the application of these concepts to the design of relational data base structures.

DEFINITION AND NOTATION

Before discussing the design of relational structures, relations (in the sense of relational data bases) should be defined and the notation to be used must be introduced.

Relations

For our purposes, a relation will be defined informally as a table (or a flat file or a rectangular array) possessing the following properties:

1. No two rows are identical
2. Row order is immaterial
3. Column order is immaterial
4. All values are atomic (single-valued)

The concept of relations is illustrated in Figure 13-1. Each column represents an attribute of a set of similar entities being described by the relation. Each row of values (called a tuple) describes an instance of the set of entities. At least one column, or one set of columns, must contain values that are unique to each row.

More formal definitions of relations are given by Codd (Reference A.8) and Date (Reference A.4).

Notation

A relation containing the attributes A, B, C, and D, with A serving as the primary key, will be denoted as shown in Figure 13-2 where the double vertical line "||" marks the demarcation between the primary key on the left and the other attributes on the right.

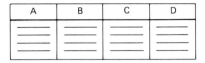

Figure 13-1. Concept of a relation.

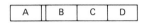

Figure 13-2. Simple primary key.

If the primary key is the composite of A and B, the relation is depicted as shown in Figure 13-3.

If all four attributes constitute the primary key, the relation is depicted as in Figure 13-4.

Fiure 13-3. Composite primary key.

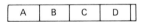

Figure 13-4. Composite primary key of all elements.

RELATIONAL KEYS

Three types of keys for a relation should be defined. The types are:

- Simple
- Fully composite
- Semicomposite

Simple Keys

A simple key is a key consisting of only one element. It is atomic and its values are unique. Examples are EMPLOYEE-NO and PART-NO.

Composite Keys

The compound keys introduced in Chapter 4 are keys composed of two or more data elements, all of which are required for unique identification. In relational terminology these are known as Composite Keys. In the relational structuring procedures, it is useful to distinguish between two varieties of composite keys.

Fully Composite Keys

A fully composite key is a key of more than one attribute so that the elements are related by an M:M mapping, a lone Type M association, or by no association at all. The key attributes are mutually independent; one is not a subqualification of the other. Fully composite keys are illustrated abstractly in Figures 13-17a and 13-17b. One example is PART-NO*SUPPLIER-NO.

Semicomposite Keys

A semicomposite key is a key of more than one attribute that is derived from an M:1 mapping (in which functional dependence is suggested). The key attributes can be considered as ordered in the sense that each key attribute (after the first) is a further qualification of its predecessors. An example is DEPART-MENT-NO*EMPLOYEE-NO and an illustration is given in Figure 13-17c.

DERIVING THIRD-NORMAL FORM RELATIONS

The data requirements, represented by the composite model (from the conceptual design), are to be structured into relations which are in third-normal form and which can be joined to provide the data paths required for the functions to be performed. To do this, the procedure must produce relations having the following characteristics:

- Normalized relations (atomic valued)   —   First-normal form
- Full-functional dependence   —   Second-normal form
- Mutually independent attributes   —   Third-normal form

Normalized Relations

A relation is normalized if each of its attributes is atomic (i.e., single-valued). The target elements of Type 1 associations are atomic by definition; the target elements of Type M associations generally are not. A Type 1 association, which represents a functional dependency (Figure 13-5a), can be structured with A as a primary key and B as an attribute in the resulting relation (Figure 13-5b).

A Type M association, on the other hand, represents a multivalued relation-

Figure 13-5a   Figure 13-5b

Figure 13-5. Normalized relation from Type 1 association.

ship (Figure 13-6a) and, in the absence of a Type 1 association in the inverse direction, must be structured with both A and B serving as a fully composite primary key to obtain unique key values (Figure 13-6b).

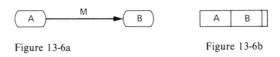

Figure 13-6a   Figure 13-6b

Figure 13-6. Normalized relation from Type M association.

Full-Functional Dependence

Full–functional dependence is obtained by ensuring that an attribute of a composite key is not also functionally dependent on a subset of that key. For example, in the situation depicted in Figure 13-7a, C is functionally dependent on A*B and also on A. The structuring rule will be to place C with its subset identifier. Thus the two relations of Figures 13-7b and 13-7c will be obtained in this case.

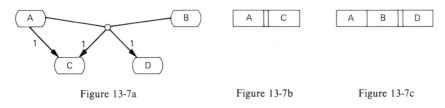

Figure 13-7a   Figure 13-7b   Figure 13-7c

Figure 13-7. Full-functional dependence.

Mutually Independent Attributes

Mutually independent attributes are assured by removing transitivity from the composite model. For example, in Figure 13-8a, C is functionally dependent on A. But C is also functionally dependent on B which, in turn, is functionally dependent on A. Thus, the association from A to C is transitive and will be removed. The relations in Figures 13-8b and 13-8c will be obtained.

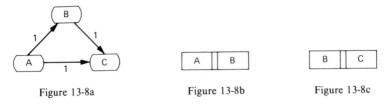

Figure 13-8a   Figure 13-8b   Figure 13-8c

Figure 13-8. Mutually independent attributes.

Implementation Comment

The test for, and removal of, transitivity will also serve to provide full-functional dependence. Consider the binary relationships of Figure 13-7a in their fully expanded form as shown in Figure 13-9a. It can be seen that transitivity exists from A*B to C, and that full-functional dependence is obtained by removing this transitive association. The result will be that shown in Figure 13-9b.

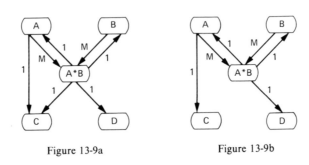

Figure 13-9a   Figure 13-9b

Figure 13-9. Obtaining full-functional dependence.

RELATIONS VS. HIERARCHIES

Many of the concepts and subtleties in the logical design of DL/I hierarchical structures do not exist in relational logical design. With relational structures, we are not concerned with such things as loops, the number of parents of a segment, nonkeyed segments, or root-only data bases. We are not dealing with floating elements needing further definition before they can be included in the structure. The relational structure, in concept, is much simpler than the hierarchical structure.

RELATIONAL STRUCTURING

The step-by-step structuring procedure is given below and is followed by an example.

Structuring Procedure

The procedure for deriving third-normal form relations from the composite model follows. Actually, this procedure appears to produce relations in fourth-normal form as defined by Fagin (Reference A.7).

1. The composite model is analyzed for functional dependency and transitivity, which are then removed. A suggested programming algorithm is given in Reference C.2.

2. Elements from which a Type 1 association emanates are primary keys. Each such key defines a relation. These can be simple keys (Figures 13-10a and 13-10c) or composite keys (Figure 13-10b).

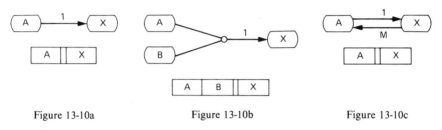

Figure 13-10a   Figure 13-10b   Figure 13-10c

Figure 13-10. Deriving keys from Type 1 associations.

3. Elements which are targets of Type 1 associations from any kind of key become attributes in that key's relation (Figure 13-11).

4. When a Type 1 association is not present between a pair of related elements, as with an M:M mapping or a lone Type M association (Figure 13-12a), the related pair is structured as a composite key. Each element pair forms a separate relation which the designer may later choose to combine. Note that lone Type M associations between keys are not augmented with inverse Type 1 associations as in the procedure for DL/I structuring. Note also, in Figure 13-12 that A, B, C, and D could have been structured into a single relation that technically conforms to third-normal form. But without information on the semantic meanings of the elements, there is no guarantee that

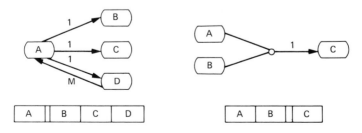

Figure 13-11. Deriving attributes from Type 1 associations.

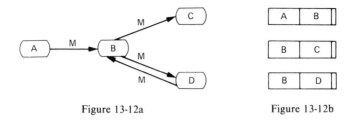

Figure 13-12a   Figure 13-12b

Figure 13-12. Deriving relations from Type M associations.

meaningful relationships exist between A and C, between A and D, or between C and D. Therefore, structuring the related pairs into separate relations (Figure 13-12b) gives the further advantages of fourth-normal form.

5. Derived relations that are subsets (projections) of other derived relations will not be included in the resulting logical model.

A Structuring Example

The composite model shown in Figure 13-13 contains the conditions of the five steps of the structuring procedure given above. The process of deriving the logical relational model is illustrated.

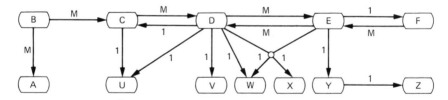

Figure 13-13. Composite model.

Removing functional dependence and transitivity results in removing the associations from D to U and from D*E to W and yields the model of Figure 13-14.

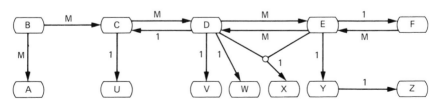

Figure 13-14. Composite model with functional dependence and transitivity removed.

Elements C, D, E, and Y are classed as simple keys because Type 1 associations emanate from them. D*E is a composite key for the same reason. With their attributes they form the relations of Figure 13-15a. It is asserted without proof that these relations constitute a minimal covering of third-normal form relations with respect to the functional dependencies (Type 1 associations) of the composite model.

Now considering element pairs related by M:M mappings and by Type M associations, the relations of Figure 13-15b are also derived. Carrying forward the minimal covering concept, Bernstein (Reference C.7) suggests combining the two left-most relations of Figure 13-15b because of the common left-hand element. But this would mean surrendering the property of Fagin's fourth-normal form (Reference A.7) and the introduction of redundant values in the extension of the resulting relation. It is recommended that this kind of combination be reserved for the designer's judgment after reviewing the results of the automated procedures as herein outlined.

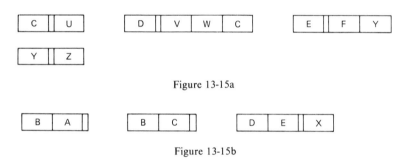

Figure 13-15a

Figure 13-15b

Figure 13-15. Relations derived from the composite model.

Note that another relation (Figure 13-16) is also derived from the M:M mapping between D and E, but since this is a subset (i.e., projection, as defined by Codd in Reference A.8) of one of the relations of Figure 13-15b, it is not included in the logical model.

Figure 13-16. Redundant relation also derived but not retained.

SPECIAL CHARACTERISTICS

The following are some special characteristics of the relational structuring process that should be noted by the reader.

Identification of Domains

A valuable item of input to the structuring procedures, over and above the binary relations of the local views, will be a list of the domains represented by the data elements and of which elements (key or attribute) belong to which domain. This information can be provided as part of the initial input or by special input after editing has taken place. Its primary use will be for determining the potential for natural joins between the derived relations.

Joins

Codd (Reference A.8) has specified certain operators that are required for manipulating relations in a data base management system. One of these operators is the JOIN in which two relations may be appropriately combined to form a new relation.

The capability for joins that support the functional requirements must be provided within the relations derived by the automated procedures. Although any two relations may be joined by computing their cross-product, this does not necessarily assure the paths needed by the functional requirements. But in the procedure herein outlined, each data element that is specified as both a source and a target element will appear in those roles in two or more derived relations. Thus the relations that must be joined to support the functional requirements will have at least one common domain (or attribute) and can be directly joined to provide the required paths.

Attributes Identified by a Composite Key

Relations derived from fully composite keys (Figures 13-17a and 13-17b) will, as expected, have the full composite key as their primary key. It is assumed from the way the binary relations are specified that all elements of the composite key are required to provide unique identification of the relation's tuples.

On the other hand, relations derived from semicomposite keys (Figure 13-17c) will have only one of the elements, as shown, as the primary key. If C is uniquely identified by A*B, and if A is uniquely identified by B, then it follows that C is uniquely identified by B. This type of situation can arise when data requirements from different local views are merged into one composite model.

Identities

If an identity exists between elements A and B (Figure 13-18a) so that each element has its own attributes, the structuring process described will produce two separate third-normal form relations keyed by A and B (Figure 13-18b).

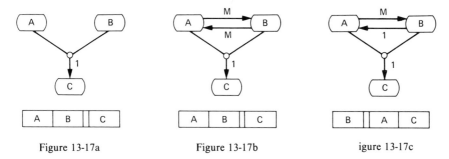

Figure 13-17a   Figure 13-17b   igure 13-17c

Figure 13-17. Deriving relations from composite keys.

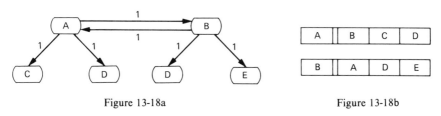

Figure 13-18a   Figure 13-18b

Figure 13-18. Resolving identities (1).

This preserves the functional dependencies in both directions between A and B.

If A and B both describe the same entity, the designers can combine A and B into a single relation without violating the intent of third-normal form.

The two elements will be assumed to describe the same entity and will automatically be assigned to a single relation if either of the following conditions exist:

- B has no (nonkey) attributes and is not related to any element other than A
- B's attributes are a subset of A's attributes

In these cases, the primary key will be the element having the most attributes as illustrated in Figures 13-19 and 13-20. Again, the functional dependencies between A and B are preserved.

All instances of identities should be reported to the designers for their review and possible restructuring.

A Second Structuring Example

As a second structuring example involving an identity, consider Bernstein's example (References C.14, C.7, C.15) as depicted in Figure 13-21.

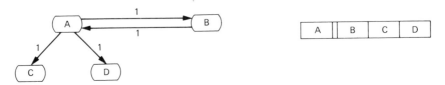

Figure 13-19. Resolving identities (2).

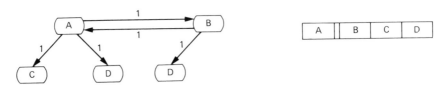

Figure 13-20. Resolving identities (3).

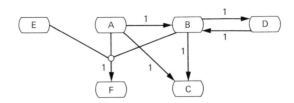

Figure 13-21. Composite model.

Removing transitivity in Figure 13-21 results in the removal of the associations from A to C and from B to A*B*E (represented by the "o"). Elements B and D constitute an identity which is resolved by considering D as an attribute of B since D has no attributes and is not an identifier of any other elements. Thus the association from D to B is also removed. The resulting composite model is as shown in Figure 13-22a, and from this the relations of Figure 13-22b are derived.

EFFICIENCY CONSIDERATIONS

Wang and Wedekind (Reference C.5), Bernstein (References C.14, C.7, and C.15), and others have made significant contributions to the concept of minimal coverings of the functional dependencies of a set of local views. This means deriving the minimal number of third-normal form relations that provide the required functional relationships. With the exception of separate relations for multivalued relationships to achieve fourth-normal form, the method outlined above will provide this minimal covering.

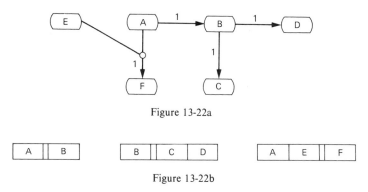

Figure 13-22a

Figure 13-22b

Figure 13-22. Deriving the relations.

The problem now becomes one of transforming this covering into an efficient covering. As stated in an earlier section about the derived DL/I segments, the third-normal form relations provide the required functional capability, but they do not necessarily provide for the most efficient storage or processing. It is sometimes advantageous to combine relations, and sometimes advantageous to split them, in order to gain storage or processing efficiencies. Information rearding frequency of use (e.g., performance weights), type of use (e.g., processing options), and expected amounts of storage space will help the designer determine the needed performance refinements.

The automated procedures that derive the canonical logical model can also produce peformance weights and relation sizes to assist the designer with efficiency considerations. With the data requirements in machine-readable form, an iterative approach with attendant quality control and documentation features (See Chapter 12), is appropriate for converging to the logical design to be implemented. The structure specification commands of Chapter 12 would, of course, have to be redesigned for relational structuring.

The main point to remember is that the minimal covering by third- (or fourth-) normal form relations constitutes a canonical logical model that serves as a point of departure for *controlled* efficiency modifications. Each modification can be made for a definite purpose and its consequences can be duly evaluated. Thus the designer can back off from the minimal covering in a controlled and intelligent manner to obtain desired efficiency trade-offs.

Performance Weights

If certain performance information can be included when defining the local views, performance weights can be calculated as part of the logical model. The general concepts for calculating performance weights were presented in Chapter 10. For relational structures, these concepts can be easily adapted to provide the following performance information.

- Frequency of use of the individual relations
- Frequency of use of the attributes within a relation
- Frequency of need for joins

Such frequencies can be useful in the performance-oriented considerations that follow.

Unnormalizing for Performance Improvement

Wang and Wedekind (Reference C.5) have given some examples of space and processing efficiencies that might be obtained by retreating from third-normal form to obtain the relations to be implemented. In particular, they cite the following instances:

1. Reducing the number of accesses by combining two or more relations frequently used by the same function into a single relation.
2. Reducing the number of accesses to update key attributes by combining relations when the key attributes of one relation are a subset of the key attributes of another relation. (But this could cause redundant storage of nonkey attributes and thus increase the update accesses for them.)
3. Reducing space requirements by splitting a relation having many tuples if the values of one attribute occupy a large number of bytes but belong to a domain of relatively few values. The attribute values are removed and placed into a separate relation and are replaced in the original relation by short pointers.

In addition to these considerations, all the reasons enumerated in Chapter 12 for refining the canonical logical design apply to relational structures as well as to DL/I.

Permanent Prejoins

For frequently invoked processing and anticipated queries, processing efficiency may be enhanced if dynamic joins can be avoided within the data base management system. This can be done by combining relations before implementation (as mentioned above) or by including the join criteria with the relational schema. In a larger sense (that also includes unanticipated queries), significant research and experimentation are being done (Reference C.8) regarding the dynamic selection of the most efficient access strategy. Hopefully, this work will lead to a reduction of the necessity for prejoins. Efficiency that can be obtained dynamically is required to give proper support to processing requirements that do change dynamically.

REPORTS

In large measure, the information content of the reports suggested in Chapter 11 applies also to relational designs. However, relational nonenclature, as well as some of the structural concepts, is quite different from that of DL/I. Therefore, with Chapter 11 serving as background material, the following is a brief summary of reports suggested for automated design of relational logical structures.

Design Reports

The following two design reports are suggested and are briefly described below:

- Derived Relations
- Suggested Joins

Derived Relations

A list of derived relations and their contents. Elements constituting the primary key will be noted. Nonkey attributes serving as entry points for one or more local views should also be noted. Characteristics of all the attributes, as on the Suggested Segments report, can be included. This report can replace the Parent-Child Graph, the Suggested Segments report, and the Candidates for Secondary Indexing report that were suggested for DL/I logical design.

Suggested Joins

A list of relations having common elements (or common domains, if that information is available). It should note which relations must be joined to support the functional requirements. A by-product would be a list of all the other natural joins that are dynamically possible. This report will replace the Candidates for Logical Relations report suggested for DL/I design.

Diagnostic Reports

The following diagnostic reports can be produced to inform the designer of choices and actions exercised by the structuring process.

Transitivity Removed

A list of transitive associations removed.

Functional Dependencies Removed

A list of associations removed to obtain functional dependence. (This list may be a subset of the Transitivity Removed list.)

Identities

A list of elements in a 1:1 mapping with each other, and how they were resolved.

Redundant Attributes

A list of data elements appearing as attributes in more than one relation.

14. GENERATING CODASYL DESIGNS

At the logical design level there are many similarities between the network structures of CODASYL and the pseudonetwork structures of DL/I. The term "pseudonetwork" is used because the logical relation and secondary indexing capabilities of DL/I transform its hierarchies into networks, although they are somewhat more restricted, as implemented, than those provided by CODASYL standards. (But the logical view, as seen by the DL/I user, is still hierarchical.) Therefore, the automated procedures thus far outlined can apply with very little modification to the logical design of CODASYL structures.

SIMILARITIES AND DIFFERENCES BETWEEN CODASYL AND DL/I

With the object of applying the automated procedures of this book to the logical design of CODASYL structures, certain similarities and differences between CODASYL and DL/I should be noted.

In DL/I, data elements are collected into segments; in CODASYL they are collected into records. Segments and records are equivalent. The segments and records may or may not contain a key field, although the automated procedures assume that a key field (possibly a dummy) will always exist. In DL/I, segments are hierarchically organized into groups in which one segment type serves as the parent of other segment types that are hierarchically subservient to it. This grouping in DL/I is analogous to the set in CODASYL, and the parent and its children in DL/I are analogous to an owner and its members in CODASYL. In pure hierarchies a child can have only one (physical) parent, but as DL/I is implemented, a child (in most cases) can have a second (logical) parent. In CODASYL, while a record can have only one owner within a set, a record can be defined in any number of sets and thus can have any number of owners. This is the fundamental difference, at the logical level, between DL/I and CODASYL.

Another major (and closely related) difference between the two logical structures is the CODASYL concept of labeled associations. In DL/I there can be only one association type in a given direction between a pair of related elements. Thus if more than one relationship exists conceptually, it must be implemented artificially and more than one alternative exists as explained below. In CODASYL it can be characterized and implemented directly by another labeled association. For example, consider the elements PLAYER and POSITION and suppose the application deals with the positions a player has played, can play, and is currently playing. In CODASYL these associations can be represented by labeled associations as shown in Figure 14-1. They can be physically implemented without redundancy as three separate sets defined on the same two record types by pointers.

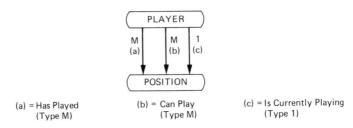

(a) = Has Played   (b) = Can Play   (c) = Is Currently Playing
    (Type M)   (Type M)   (Type 1)

Figure 14-1. Concept of labeled associations.

These same associations could be represented in DL/I (assuming PLAYER and POSITION are both keys) (1) by a single POSITION segment with a flag field to indicate the nature of the relationship, (2) by separate POSITION segments for each type of relationship, or (3) by logical relations between the PLAYER segment type and a POSITION segment type. As was explained in Chapter 4, the resulting DL/I implementation depends on how the designer specifies the data requirements. For CODASYL, the designer is not concerned with alternate implementation structures. He merely specifies the labeled associations between the pair of elements.

STRUCTURING CODASYL LOGICAL DESIGNS

In large measure, the structuring procedures outlined in the preceding chapters can apply directly to the design of CODASYL logical structures. But there are some noteworthy exceptions.

Since, in a set, a member can have only one owner and that owner can be a member of some other set, the procedure of following Type 1 associations to determine segment contents and child-to-parent segment relationships (See Chapter 6) also applies in deriving record contents and member-to-owner relationships. The concept of augmenting lone Type M associations between keys by supplying inverse Type 1 associations can also apply for helping to define sets. The same subtleties (See Chapter 8) apply and the same intelligence (See Chapter 9) can be built into the structuring procedures. The primary difference is that in CODASYL a record can be a member of any number of sets. Thus, instead of trying to determine the one physical parent from a group of candidates as is done for the DL/I structuring, all candidate parents can be included in the resulting logical model as owners of their own sets. In addition, the occurrences of loops, which should still be reported for designer information, does not necessarily represent structural incompatibilities in CODASYL.

The procedures outlined in Chapter 7 for determining candidates for logical relations and secondary indexes also apply to CODASYL structuring. The M:M mappings and the Type M associations which led to physical parent-logical parent relationships, and which required a third segment to define the intersection, now lead to two (or more) owner record types requiring a link record (which may or may not contain intersection data). And the criteria for determining candidates for secondary indexing now lead to candidates for system segments permitting ordered entry into the network at various alternate locations.

The performance weight calculations of Chapter 10 can apply in totality to CODASYL designs to serve as one of the criteria for refining the canonical (initial) structure produced by the automated procedures. And the same iterative approaches for editing and for structuring, by gradually adding additional requirements to that which has been structured for the more important requirements, are advocated and made possible because of the automated techniques. The concept of structure specification commands suggested in Chapter 12 still applies generally, but they will have to be specifically designed for CODASYL structuring.

As an example of deriving the logical design of a CODASYL structure, consider the six local views of Figure 14-2. Assume that all data elements shown are to be treated as keys and that attributes, not shown, are specified for each of them.

Combining these local views (editing is not illustrated in this example) yields the composite model of Figure 14-3. But assume that as a result of editing, the designers decided that the Type 1 association from PLAYER to POSITION in the local view of Figure 14-2b should be labeled "Is Currently Playing."

Records and their contents are first determined by identifying keys and attributes. Keys are elements that identify other elements with Type 1 associations. Attributes are identified by keys, but do not, in turn, uniquely identify other elements. (Remember that we are considering all elements in this example to be keys.) Sets are derived by examining M:1 mappings or lone Type M associations, and designating the "from" element of the Type M association as the owner and the "to" element as the member.

Alternately, the procedures outlined in Chapter 7 for deducing physical child-to-parent relationships by following Type 1 associations upward between keys can be applied directly to the derivation of member-to-owner relationships in a bottom-up process. This approach requires augmenting the lone Type M associations with Type 1 inverses (as was done for DL/I structuring). Figure 14-4 illustrates the example of Figure 14-3 with the Type 1 inverse associations added.

Regardless of the alternative used, the designer should be provided with a

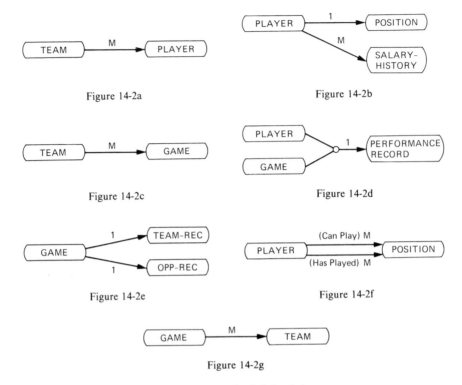

Figure 14-2a

Figure 14-2b

Figure 14-2c

Figure 14-2d

Figure 14-2e

Figure 14-2f

Figure 14-2g

Figure 14-2. Example of six local views.

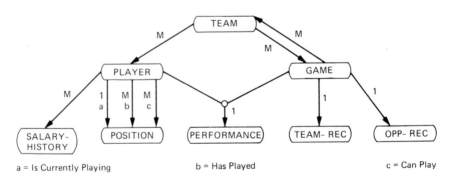

a = Is Currently Playing   b = Has Played   c = Can Play

Figure 14-3. Resulting composite model.

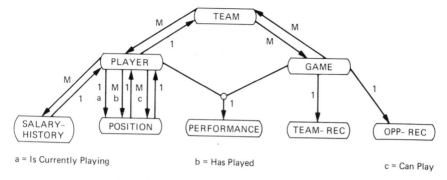

a = Is Currently Playing   b = Has Played   c = Can Play

Figure 14-4. Composite model with inverse associations added.

list of lone Type M associations so he can evaluate the resulting structuring. And he must have means for overriding the derived result.

An M:M mapping between a pair of elements translates into two record types to be related by a LINK record or by a record containing intersection data.

Finally, there is no concept of inconsistent associations with CODASYL structures. If there are two association types in the same direction between a pair of elements (as is the case from PLAYER to POSITION), Type M takes precedence in establishing the existence of a set. The Type 1 association can be implemented by including a flag field in the POSITION records to indicate which position is intended, or it can be implemented by a separate pointer, thus establishing an additional set on the same record types.

Now by applying the systematic logical design procedures, as adapted to CODASYL structuring, the logical model of Figure 14-5 will be obtained. Note that because PLAYER is the root of one (or more) of the local views but is not a data base root, it is structured into a singular set containing SYSTEM as the owner. This suggests an indexed entry into the data base through PLAYER.

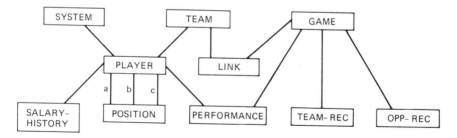

Figure 14-5. Derived logical model.

REPORTS

This section will also be presented on an exception basis. With the exceptions to be noted below, the reports suggested in Chapter 11 can also apply to CODASYL logical design.

Design Reports

The primary exception in the design reports is that the Parent-Child Graph will depict a network rather than a series of hierarchical trees. As such it can be called the Network Graph.

A Link Record Summary and a System Record Summary can replace the Candidates for Logical Relations and the Candidates for Secondary Indexing reports. These reports, while not as critical as they might be for DL/I structuring, can alert the designer to all places where the use of links and system records should be evaluated.

Diagnostic Reports

With regard to the diagnostic reports of Chapter 11, the path rule violations can be renamed the Loop Report, and the reporting of paths of more than 15 elements (which is now meaningless) can be dropped. Repeating attributes, which are a violation of the approach taken for the DL/I structuring and which are a violation of the atomism of relational structuring and which are a violation of the atomism of relational structures, are very much a part of CODASYL designs. This type of report, however, may still be of interest to the designer in his editing considerations.

PART IV
PHYSICAL DESIGN

Physical data base design means augmenting the logical model with the choices necessary for defining the physical storage and usage options of the data base and then evaluating the result in terms of required storage space and processing efficiency. The choices and options include how and where the data is to be stored, how it will be located, and how it will be used. Specifically, for DL/I data bases, we mean making those choices that must be specified in the DBD control blocks.

We will be interested in two types of evaluation: storage space and I/O performance. Three major factors of I/O performance are I/O time, central processing unit (CPU) time, and resource contention. Any useful performance evaluation must provide comparative information in these areas.

The logical design procedures that have been presented provide, in addition to a suggested logical design, alternatives for the designer's consideration. Therefore, the designer may have, in effect, a family of candidate logical designs from which to choose. The physical design process is frequently required to fully evaluate the performance implications of some of the individual candidate logical designs. Analytic calculations are a convenient means of performing a gross evaluation to quickly reduce the number of candidate designs. Final evaluation may require the more precise methods of modeling.

Analytic calculations will be presented for estimating storage space. Additional analytic estimates will be developed for evaluating I/O time and CPU time, but they ignore the issue of resource contention. Modeling techniques will be presented for use when more accurate performance information, based on all three criteria, is required.

Computer assistance can make a major contribution to the physical design process simply by providing an online conversational dialogue which can guide the designer in making and specifying the physical design choices and options.

Starting with the hierarchical structure, segment content, segment lengths, frequencies of occurrence, frequencies of use, and processing options (all of which can be available from the automated logical design techniques), a conversational terminal session can guide the designer in selecting device type, access methods, block sizes, secondary data set groups, pointer options, root anchor points, logical relationships, and indexing and/or randomizing choices.

In addition to guiding the designer in these choices, automated techniques can also be used to make helpful suggestions, such as recommending the most efficient access method for the processing to be performed, calculating optimum block sizes, evaluating alternate randomizing techniques, and presenting information helpful in determining secondary data set groups. Again, it is emphasized that such suggestions are to be reviewed and possibly overridden by the designer.

In the following chapters we present techniques and equations for computer assistance in physical design. We will also present guidelines for many of the choices that must be made by the designer. The treatment will be addressed exclusively to DL/I data structures.

To assist with understanding what follows, Chapter 15 presents a brief review of the basic DL/I access methods. Readers already familiar with this topic may choose to skip this chapter. Chapter 16 treats the choices to be made in order to produce Data Base Descriptions (DBDs) and suggests specific types of computer assistance. Chapter 17 develops equations for estimating required storage space and optimum block sizes and Chapter 18 derives I/O probabilities and analytic I/O timing estimates. Chapter 19 addresses application program modeling for detailed performance evaluation of both the data base design and the application program's DL/I call sequences. Finally, Chapter 20 presents guidelines and overall considerations for practical aspects of physical design.

15. Brief Access Method Review

The techniques and guidelines of physical design of DL/I data bases are well known among practitioners and are well documented in the literature; hence, they will not be repeated in detail here. However, as a tutorial for those not familiar with these procedures, and also to serve as points of reference for the further suggestions to be made about computer assistance, a brief review of the storage patterns and processing characteristics of the basic DL/I access methods is in order. The treatment will be cursory; its purpose is merely to set the stage for the considerations of access method selection and optimum storage space calculation. More comprehensive treatments are provided in Date, Chapter 16 (Reference A.4), Kapp and Leben, Chapter 12 (Reference D.1), McElreath, Chapters 11 and 12 (Reference A.6), and the IMS/VS System/ Application Design Guide, Chapter 4 (Reference D.5).

ACCESS METHOD OVERVIEW

For DL/I data bases there are two basic storage organizations with two variations of each, resulting in four basic DL/I access methods. In the *Hierarchical Sequential* (HS) storage organization, segments within a data base record are accessed sequentially according to their hierarchical sequence which is reflected in their relative physical positioning. The Hierarchical Sequential Access Method (HSAM) accesses the root segments sequentially from the end of the previous record. It is a sequential access method in the traditional sense because it always starts at the beginning of the data base and works its way sequentially segment by segment and record by record. The Hierarchical Indexed Sequential Access Method (HISAM) is similar in concept to HSAM, the primary difference being that in HISAM, access to any root segment may be made directly via an index. Except for pointers to and from overflow blocks, segment sequence is again determined by physical positioning on the storage device.

The *Hierarchical Direct* (HD) storage organization provides direct addressing to root segments and direct addressing also from any segment to its next twin, to the first occurrence of its immediate child segments, and to its parent. Direct accessing for hierarchical sequence is also available. The primary differences between the two HD access methods are in the method of locating the root segments. In the Hierarchical Direct Access Method (HDAM), root segments are addressed by means of a randomizing routine that determines the physical location of the root by calculation based on the value of the root's key. In the Hierarchical Indexed Direct Access Method (HIDAM) accesses to root segments are via an index.

Although other DL/I access methods, such as General Sequential Access Method (GSAM), Simple Hierarchical Sequential Access Method (SHSAM), and Simple Hierarchical Indexed Sequential Access Method (SHISAM) are available, our discussion will be limited to the four basic ones introduced above. They are the ones of major importance; the others are used to access nonhierarchical sequential structures in a DL/I environment.

The four basic DL/I access methods—HSAM, HISAM, HDAM, and HIDAM—provide interfaces between the application programs and the DL/I data base management system. But these DL/I access methods are actually built upon the operating system access methods: Basic Sequential Access Method (BSAM), Queued Sequential Access Method (QSAM), Indexed Sequential Access Method (ISAM), Virtual Sequential Access Method (VSAM)—both Keyed Sequenced Data Set (KSDS) and Entry Sequenced Data Set (ESDS)—and upon Overflow Sequential Access Method (OSAM) which is provided by IMS and by DOS-DL/I. The storage organizations of the four basic DL/I access methods are depicted in Figure 15-1, and

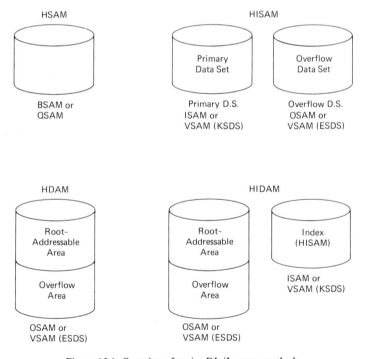

Figure 15-1. Overview of major DL/I access methods.

	BSAM	QSAM	ISAM	VSAM (KSDS)	OSAM	VSAM (ESDS)
HSAM	x	x				
HISAM						
Primary			x	x		
Overflow					x	x
HDAM						
Root-Addr					x	x
Overflow					x	x
HIDAM						
Root-Addr					x	x
Overflow					x	x
Index			x	x		

Figure 15-2. Composition of major DL/I access methods.

their relationships to the system access methods, and to OSAM, are summarized in Figure 15-2.

GENERAL CHARACTERISTICS AND STORAGE PATTERNS

Segments are organized into blocks which are the units of information transferred across channels by an I/O operation. With VSAM, the control interval is the block. The storage patterns, which vary slightly according to the access method being used, have significant implications with respect to the storage space required and to the performance characteristics of the resulting data base.

In order to provide examples of representative storage patterns for the basic DL/I access methods, the hierarchical structure of the example shown in Figure 15-3 will be used. We will assume fixed length segments with the length

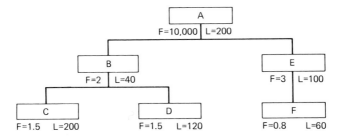

Figure 15-3. Example of an average data base record.

including both the data and the prefix. For each segment type, the expected frequency of occurrence (F) and length (L) are as indicated.

A data base record consists of a single occurrence of a root segment plus all occurrences of the segment types hierarchically subordinate to that root. For the logical structure defined in Figure 15-3, two example data base records are illustrated in Figures 15-4 and 15-5. Segment type lengths are in parentheses. These two example records will be the basis for the storage patterns that follow.

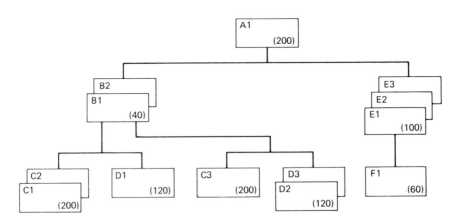

Figure 15-4. Example of a specific data base record (1).

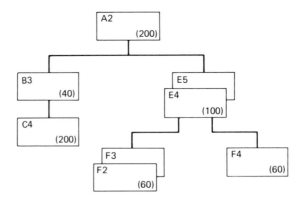

Figure 15-5. Example of a specific data base record (2).

The following is a summary of the storage patterns of the four basic DL/I access methods. A more detailed presentation may be found in Reference D.5.

Hierarchical Sequential Access Method (HSAM)

The HSAM data base is a series of blocked segments loaded in hierarchical sequence. Since updates are not permitted, this sequence does not change. However, updates can be made, as with tape files, by copying the data base.

HSAM blocks are fixed length with no distinction between primary and overflow blocks. Assuming block lengths of 800 bytes, the two data base records illustrated in Figures 15-4 and 15-5 would be blocked by HSAM as shown in Figure 15-6.

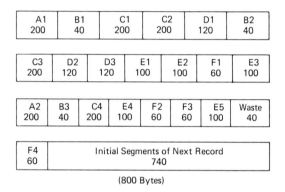

Figure 15-6. Example of HSAM blocking.

Hierarchical Indexed Sequential Access Method (HISAM)

The HISAM data base is organized into two distinct data sets: primary and overflow. Segments are organized into logical records which in turn are blocked. Each root segment begins a new logical record which contains as many dependent segments (in hierarchical sequence) as will fit. When loading or reorganizing, all root segment logical records are in the primary data set. For each data base record, those dependent segments that do not fit into the primary (root segment) logical record are placed into overflow logical records which are blocked into the overflow data set. These overflow logical records are chained to the primary logical records in a manner that preserves the hierarchical sequence of the dependent segments.

The method of inserting root segments depends on whether VSAM or ISAM/OSAM is being used. Using VSAM, root segments are inserted into the primary data base so that they are physically positioned according to key value sequence. Within the appropriate block, if one or more empty logical records exist, then filled logical records are shifted one place to the right to make room for the new root segment in its proper relative position. If all logical records in the block are already filled, the insertion is preceded by a control interval split in which the right-most half of the logical records are placed into a new control interval.

Using ISAM/OSAM, a root segment is inserted into the overflow data set into the next available logical record, and it is appropriately chained to the primary logical records to preserve key sequence order.

Dependent segments are inserted into the appropriate logical record (primary or overflow) and maintain hierarchical sequence by physical positioning, provided space is available in the logical record. This is done by physically shifting segments to the right within the logical record to make room in the appropriate place. If space is not available, the right-most segments (and possibly the inserted segment itself) are placed into one or two overflow logical records, and hierarchical sequence is maintained by appropriate chaining.

Deleted segments are not physically removed, and their space is not available for reuse until reorganization.

To illustrate the HISAM storage patterns, we will assume a logical record size of 600 bytes for both the primary and the overflow data sets. This is an unrealistically small size, but it will work well for illustrative purposes.

Consider the data base record in Figure 15-4—it consists of root segment A1 and the 12 occurrences of its dependent segments. The HISAM data base will have the logical record pattern with segment lengths as indicated in Figure 15-7.

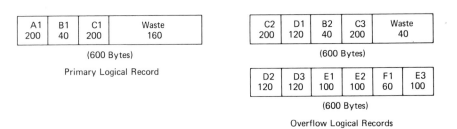

Figure 15-7. Example of HISAM logical records (1).

Notice that one of the overflow logical records is completely filled with no waste space; this is the exception rather than the rule. In the first overflow logical record, 40 out of 600 bytes are wasted because the next segment in hierarchical sequence will not fit. This is not an unreasonable amount of waste. But in the primary logical record, 160 of 600 bytes are wasted. A better choice of logical record size might result in reduced waste space.

The concept of waste space is further illustrated by considering the next sample data base record (see Figure 15-4) with root segment A2 and seven

A2	B3	C4	E4	F2	Waste
200	40	200	100	60	40

(600 Bytes)

Primary Logical Record

F3	E5	F4	Available
60	100	60	380

(600 Bytes)

Overflow Logical Record

Figure 15-8. Example of HISAM logical records (2).

occurrences of its dependent segments. For this data base record, the logical records will be loaded as shown in Figure 15-8.

For this second data base record, the waste space in the primary logical record is not excessive. But the 380 unused bytes in the overflow logical record illustrate a second kind of waste space—there is more space in the logical record than needed for the segments that are to be placed inside. This kind of waste can be filled during processing by segment insertions and thus can be considered as free space, to allow for expansion, rather than as true waste space. Thus the only waste space we will consider in the space calculations of Chapter 17 will be the space that is left by segments that do not fit.

So far we have considered only two data base records. The real question is how suitable the selected logical record size is for the data base as a whole. One of the major physical design considerations for HISAM data bases is the selection of optimal logical record sizes for minimizing waste space. We shall give special attention to this consideration in Chapter 16.

While it is true that the HISAM logical records can be blocked, block size is chosen as an even multiple of logical records so there is no additional waste space in blocking. In general, when using ISAM, HISAM logical records are blocked as in Figure 15-9.

L.R. 1	L.R. 2	L.R. 3	· · ·	L.R. n

Figure 15-9. Concept of logical record blocking.

Assuming a blocking factor of two for the example at hand, the HISAM logical records thus far derived will have the storage patterns of Figure 15-10.

The primary blocks are contained in the HISAM primary data base and the overflow blocks are in the HISAM overflow data base.

Hierarchical Direct Access Method (HDAM) and Hierarchical Indexed Direct Access Method (HIDAM)

The HDAM and HIDAM storage patterns are essentially the same, so we will speak generally of the HD organization storage patterns. The HD organization

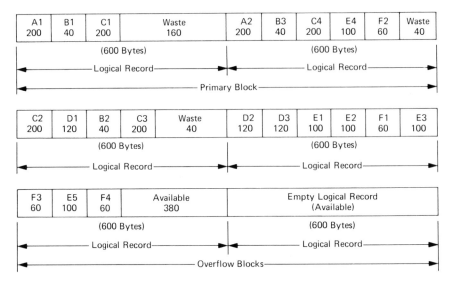

Figure 15-10. Example of HISAM blocking.

has only one data set (ignoring grouping) that is divided into a root–address-able area and an overflow area. The root segment and its subordinate segments (in hierarchical order) up to a specified number of bytes are loaded into root-addressable blocks. Additional segments of the data base record are blocked into the overflow area. Because a block cannot end with a partial segment, there is usually some waste space at the end of each block.

In HDAM, roots are accessed by a key randomizing technique. Root anchor points are devices used to enhance the ability of the randomizing routine to generate a unique physical address for each key. Root anchor points subdivide the root-addressable area blocks into a specified number of addressable slots. The randomizer uses the key value of the root segment to generate the physical address of a block and a slot (i.e., root anchor point) within the block. In HIDAM, roots are accessed via an index and root anchor points are not used.

Segments are not moved to make room for insertions. To allow for data base expansion, the designer can specify a number of blocks to remain free at load time and an amount of space within each block to remain free. These free spaces are important considerations when calculating data base storage space.

Insertions of root segments are made to the root–addressable area as long as there is sufficient space. Otherwise, they are made to the overflow area. When inserting dependent segments, an attempt is made to place them as close to their roots as possible while at the same time minimizing the fragmentation of free space. They will be placed into the root–addressable area if the byte limit permits; otherwise, they will go into the overflow area.

Space from deleted segments is theoretically available for immediate reuse, although this is not always the case in actual practice.

In general, the HD blocks are organized as shown in Figure 15-11. Some waste space normally exists at the end of an HD block because of lack of space for the next segment to be loaded.

Root-Addressable Area Block

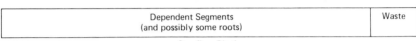

Overflow Block

Figure 15-11. Concept of HDAM and HIDAM blocking.

To illustrate the HD storage patterns and the concept of waste space within a block, we will assume a root-addressable area byte limit of 300 bytes and an effective block size of 900 bytes (10% free space in a 1,024 byte block). Ignoring system overhead fields, the example data base records of Figures 15-4 and 15-5 will yield the storage patterns shown in Figure 15-12.

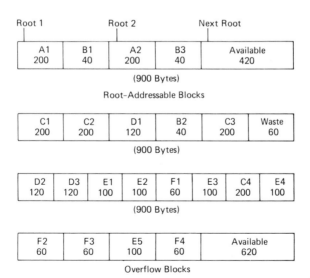

Figure 15-12. Example of HDAM and HIDAM blocking.

The root-addressable block is loaded as full as the byte limit rules will permit although there is room for a root and dependent segments from the next data base record. The first overflow block has 60 bytes waste because the next segment, D2, would not fit. The second overflow block is filled to capacity and the third has space available for additional dependent segments.

PERFORMANCE CHARACTERISTICS

Generally speaking, the HD organization is best suited for volatile data bases which are subject to frequent insertions and deletions. It is also used for relatively stable data bases in which high-performance nonsequential accesses are required. The HS organization is usually preferred for stable data bases in which segments within a record are to be processed sequentially.

When initially loaded or after being reorganized, the HS organization may perform better than the HD organization for sequential and skip-sequential processing. But in the HS organization (i.e., in HISAM), insertions and deletions cause segments to be physically moved on the storage device, and insertions cause chains of segments to be placed into an overflow area. Thus, as such updates proceed, the sequential ordering of the HS segments becomes compromised, performance degrades, and the HD organization becomes the better performer. As a corollary, HS organizations (HISAM) usually require reorganization more frequently than HD organizations.

The following summary of relative advantages and disadvantages of each access method is derived primarily from *The Data Base Design Guide* (Reference D.3).

HSAM

Advantages

1. The only DL/I access method supporting tape devices.
2. Conversion from Operating System (OS) data sets is relatively easy.

Disadvantages

1. Must rewrite file to update.
2. Logical relationships are not available.

HISAM

Advantages

1. Sequential retrieval without using the index can be relatively fast if the file is well organized.
2. Provides direct access via an index to root segments of a sequential file.
3. Faster loading than HIDAM because random arm movement is not required.

Disadvantages

1. Using ISAM/OSAM all inserts go into the overflow data base, although with VSAM, root segment inserts go into the primary data base.
2. No direct pointers to the root or to the other segments.
3. Segments are physically moved to make room for inserts.
4. Deleted segment space not immediately available for reuse.
5. Logical relationships are somewhat restricted (e.g., virtual pairing).

HDAM

Advantages

1. Can have nonunique root segment keys.
2. Space from deleted segments can normally be reused immediately.
3. Segments are not physically moved to make room for inserted segments.
4. A data base may be split into data set groups at any hierarchical level.
5. Generally faster random accessing than with an index.

Disadvantages

1. No efficient sequential accessing of root segments. (They can be accessed sequentially, but their keys will be in random order according to their random distribution on the storage device.)
2. Accessing of root segments can be slower than with an index if the randomizing algorithm is not well suited for the distribution of key values or if root segments are not in their home addresses.

HIDAM

Advantages

1. The index may be reorganized without reorganizing the indexed data.
2. Space from deleted segments can normally be reused immediately.
3. Segments are not physically moved to make room for inserted segments.
4. A data base may be split into data set groups at any hierarchical level.

Disadvantages

1. Slower loading than HISAM. Loading time may be improved by sorting the records into their randomized physical address sequence.
2. Direct access to a root segment may be slower than HDAM because the index must be accessed.

16. Physical Design Choices and Options

This chapter treats certain choices and options needed to convert the logical model into a physical model ready for storage allocation and loading. More specifically, it addresses a selection of DBD considerations where computer assistance appears to have special application.

PHYSICAL DESIGN CONSIDERATIONS

There are many choices to be made among the physical design parameters. Access method is one of the major considerations because most of the other considerations depend heavily on the access method chosen. Since our purpose here is to suggest areas where computer assistance can be helpful and not to provide a manual on physical design, several of these considerations are listed in general terms rather than in specifics. For example, logical relations are listed as a general consideration. Treating the details of unidirectional or bidirectional relationships, physical or virtual pairing, location of destination parents, etc., is beyond the intended scope of this chapter. Similarly, details such as the number of root anchor points, size of the root-addressable area, etc., are not delineated for HDAM randomizing and blocking although they will be mentioned generally in the text to follow.

The general physical design considerations, roughly in the order in which they must be addressed, are enumerated in Figure 16-1.

Those areas that are to be individually addressed in this chapter are flagged with an asterisk. An online conversational procedure can give general assistance by guiding the designer through all these choices, and can give specific help in some of the areas. Some choices, in real life, are usually predetermined. For example, the choice of device type is usually based on general price-performance criteria of the device itself and has often been determined before designing the data base. Decision tables can be the basis of assisting the designer in selecting access methods, pointer options, insert-replace-delete rules, etc. Other choices such as determining unidirectional or bidirectional logical relationships and virtual or physical pairing can be significantly aided by the structures and performance information from the logical design process.

Without attempting to cover the entire physical design spectrum, certain

P. D. Consideration	HSAM	HISAM	HDAM	HIDAM
Device Type	*	*	*	*
* Access Method	*	*	*	*
* Block/Control Interval Size	*	*	*	*
* Logical Record Size		*		
Indexing		*		*
* Randomizing			*	
Logical Relations		*	*	*
Pointer Options			*	*
* Secondary Data Set Groups		*	*	*
Secondary Indexes		*	*	*
Insert-Replace-Delete Rules		*	*	*

Figure 16-1. Physical design considerations.

areas will now be explored to illustrate the kinds of computer assistance available.

ACCESS METHOD SELECTION

By considering the characteristics of the processing to be done on a given data base, it is possible to remove much of the guesswork from DL/I access method selection. Certain combinations of processing characteristics suggest certain access methods as being most appropriate. Thus, a decision table approach to access method selection is suggested; a number of such approaches have been developed. The decision table of Figure 16-2 was developed by the GUIDE IMS State of the Art Project (Reference D.3).

In order to understand the decision table properly, we must be familiar with the performance characteristics of the access methods presented in Chapter 15. Based on the processing characteristics of the access methods as enumerated, decision tables like Figure 16-2 can be derived to assist the designer in selecting the "best" access method for a given processing situation. In this type of approach, an interactive conversational dialogue at a computer terminal can help the designer provide the necessary conditions for the decision table's stubs. Note that the decision table also makes suggestions about pointer options and data set groups.

Condition				
All processing is batch seq.; high activity on file	YYYYY			
Or processing largely seq.; some indiv. query		YYYYYYYYY		
Or only file maint. is seq.; only high activity is update			YYYYYYYYYYYY	
Or partly seq.; high query and trans. activity				YYYYYYYYYYYY
And has a future potential as a data base	Y	Y	Y	Y
Has a high degree of redundant data across files	YY	Y	Y	Y
Data fields added, descriptions changed frequently	Y	Y	Y	Y
Need to access data on a variety of different keys		Y	Y	Y
Need fast response or low volume, unpredictable incidence of queries		Y	Y Y Y Y	Y Y Y Y
Need to establish relationships between segs. of the same data base		Y	YY	YY
Need to establish relationships among several data bases		Y	YY	YY
High activity on a portion of the data base, low activity on the remainder	Y	Y	YY	YY
Traditional Seq.	X X		X	

Figure 16-2. Access method selection decision table. (Continued on p. 152)

Sequential DL/I HSAM	X X X X		X X		
Indexed Sequential DL/I HISAM	XX X		X		
Direct Address DL/I HDAM			X X X X		X X X X
Indexed Direct Addr DL/I HIDAM		XXXXX		X X X X XXXXX	X X X
And Physical Ptrs.		XX		XXXX	XXX X
Logical Pointers		X		XX	XX
Mult. DS Groups	X	X		XX	XX

Figure 16-2. (Continued)

OPTIMUM BLOCK SIZES

In discussing HISAM, the logical record size was considered the basic unit for storage space considerations. In this chapter and in Chapter 17 we will use: "block" as the basic unit for storage space considerations. Roughly equivalent terms are "control interval" for VSAM access methods, and "logical record" for HISAM based on ISAM. The concepts to be presented apply to all these cases.

The consideration of block size can involve several factors. For example, the choice of block size is influenced by the type of storage device used. It also affects the choice and use of the system's buffer pools. Block sizes that are too large can mean transferring unneeded information across the channels and usurping buffer pool space; block sizes that are too small can mean extra I/Os because of not transferring enough data. Perhaps the major concern about block sizes is the resulting storage space required for the data base, and this is the aspect we shall discuss.

We are interested in choosing the block size that will minimize the overall data base size. Smaller blocks tend to have a higher percentage of waste space caused by segments not fitting at the end of the block. Since small blocks mean more blocks, the waste problem is compounded. At the other extreme, large blocks that can contain entire data base records will have unused space when the record is shorter than the block. Figure 16-3 depicts the typical relationship between block size and required storage space.

For beginning guidelines, blocks must be at least as large as the largest segment in the data base and should be no larger than the largest data base record.

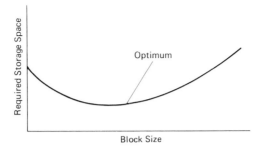

Figure 16-3. Concept of optimum block size.

In most cases, both these extremes still imply inefficient use of space, and the most efficient block size is somewhere in between. In Chapter 17 we will develop equations for estimating required storage space with block size being one of the parameters. Taking this a step further, we are tempted to seek the optimum block size by differentiating the space equation with respect to block size and setting the result equal to zero. If the functional relationship were smooth and continuous, this would be an analytic way of determining the block size that minimizes storage space. But such a continuous relationship does not necessarily exist. In addition, when using VSAM, there are only certain eligible values of block size; they are multiples of 512 bytes up to a point and of 1,024 bytes after that. An analytically calculated block size does not necessarily give a clue to the best of the eligible block sizes.

Computer assistance can be employed to determine optimum block size simply by calculating required storage space for each of the candidate block sizes and by presenting the results in a graphical form similar to that of Figure 16-3. If secondary data set groups are to be used, block size should be calculated for each group within the constraints of the access methods being used.

DATA SET GROUPS

Except for the HSAM access method, a data base may be divided into secondary data set groups according to the rules given in the *IMS/VS Version 1 System/Application Design Guide* (Reference D.5) and in the *IMS/VS Utilities Reference Manual* (Reference D.6).

There are several reasons for using data set groups. By providing a way of separating frequently used data from infrequently used data, groups can contribute to improved performance by shortening the search path from a current segment to the segment next to be accessed. For HISAM, this can reduce the number of physical I/Os by reducing the number of blocks to be searched. For both HISAM and the HD organizations, it also tends to reduce physical I/Os (the number of blocks actually transferred to or from secondary storage) by increasing the likelihood that the desired segment is in the same block as the current segment. For HISAM data bases, groups also give an additional performance advantage by enabling direct accessing of segment types at the second level without first going through the root. Performance weights, segment sizes, and frequencies of occurrence can all be made available in convenient form by the logical design procedures to aid in such considerations.

Another function of groups is to reduce the online storage requirements of some large data bases by subdividing them for periodic processing. This explains why some companies process billings for some customers in one week of the month and for other customers during another week of the month. For data bases in which some segments are significantly more volatile than others, data set groups provide a means of loading the data base so that the more volatile segments are less densely packed than the less volatile segments because each data set group can have its own free space characteristics.

A fifth reason for using data set groups is to provide more efficient use of storage space by grouping segment types of significantly different sizes into different groups. Optimum block sizes can be individually specified for each group so that waste space is further reduced. Both graphic presentations of segment sizes (Figure 16-4) and total expected segment bytes (Figure 16-5)— which are derived from the average data base record of Figure 15-3—can be helpful input to the designer when considering this type of grouping. This infor-

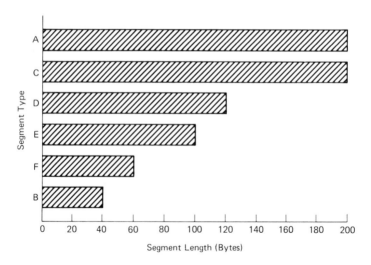

Figure 16-4. Display of average segment lengths.

mation can be provided with virtually no additional effort by the automated logical design procedures. Based on the consideration of storage space alone, the use of such information is illustrated in the following paragraphs.

While the variation of segment sizes is important in determining grouping requirements, an estimate of the total number of bytes required by each segment is also needed. A segment type of a "different" size having only a few occurrences may not justify creating a separate data set group. The total number of bytes for a segment is estimated by multiplying the segment length by the product of the frequencies of all the segments in the path from the root down to and including the segment itself. For the example being illustrated, the graph shown in Figure 16-5 can be derived.

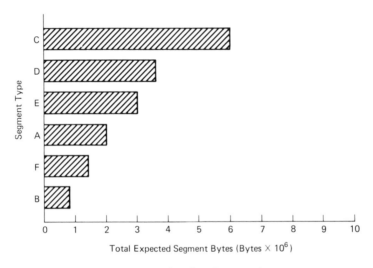

Figure 16-5. Display of total segment bytes.

These graphs show C and the root segment, A, to be the same size, so we would like to keep C in the primary group with A. But B, which is hierarchically between A and C, is the shortest segment type, being one-fifth the length of A and C. Since the total expected bytes for B is also small compared to the combined lengths of A and C, there is not an overwhelming demand to separate B from A and C. It seems reasonable to retain B in the primary data set group with A and C. The treatment of D and E depends on whether the access method is HISAM or whether it is HDAM or HIDAM.

With HISAM, secondary data set groups must begin with a second level segment type and contain all hierarchically dependent segment types. Therefore, D must remain in the primary group with B, and E and F must be together. E and F are much shorter than A and C, and since the total expected bytes for E (three million) is relatively significant, E and F can be placed into a secondary data set group. The resulting grouping (Figure 16-6) is not unreasonable.

HISAM Primary Data Set Group 1	HISAM Secondary Data Set Group 2
A, B, C, D	E, F

Figure 16-6. Possible data set grouping for HISAM.

We have more flexibility when using HDAM or HIDAM; segments can be grouped in any combination. Looking only at segment lengths, the natural groupings seem to be A and C, D and E, and B and F. Because B still occupies a relatively small total number of bytes, we will keep it with A and C in order to avoid additional I/Os when going from A to C. But it appears reasonable to group D and E together because of their similar sizes and to let F, which is one-half of the size of D, constitute a third group. This possible result is depicted in Figure 16-7. But before reaching a final conclusion, the frequency with which D is accessed from B could determine the inclusion of D in the primary group with B.

HD Primary Data Set Group1	HD Secondary Data Set Group 2	HD Secondary Data Set Group 3
A, B, C	D, E	F

Figure 16-7. Possible data set grouping for HDAM or HIDAM.

We are not suggesting that the selection of data set groups can or should be an automatic process. At the present state of the art, it still needs human judgment. But we do suggest that as a by-product of the logical design procedures, information regarding segment sizes, frequency of use, and type of use can be produced in a form that can be very helpful to the designer in his grouping considerations.

RANDOMIZER EVALUATION

An ideal randomizing (or hashing) algorithm is one that transforms every root segment key into a unique physical storage address so that the roots are uniformly distributed in the allocated storage area. Ideal randomizing algorithms generally do not exist; in actual practice collisions and poor distributions tend to occur. Collisions take place when two or more root segment keys (i.e., synonyms) are randomized to the same physical address. In such cases, the first root segment goes into its "home" address with the synonyms being placed as near to it as possible and chained serially to the home location. Two problems result from synonyms. Additional I/O time is required to follow the chains, especially if the synonyms are in different blocks, and the synonyms occupy space that could be the home address of root segments yet to come. Clustering is also a problem with randomizing. Some portions of the storage area may be densely filled while other portions remain relatively empty. Again, densely filled areas may have little or no room for additional root segments or for insertions of dependent segments.

Most randomizing algorithms yield good results while the data base is lightly loaded. But as the packing factor increases (i.e., as the data base becomes filled), the effectiveness of the randomizing algorithm tends to deteriorate according to the pattern illustrated in Figure 16-8. Thus, for a given distribution of key values, at what packing factor does each randomizing routine cease to be effective? In other words, which of a set of candidate randomizing algorithms will yield the best results at high packing factors?

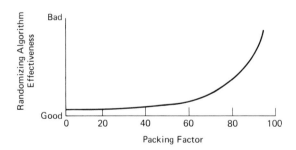

Figure 16-8. Concept of randomizer efficiency.

Computer assistance can be very helpful in evaluating candidate randomizing algorithms. If the expected distributions of the root key values can be ascertained or approximated, then the expected distribution patterns of root segment placement can be calculated and studied for various packing factors and DBD parameters. For example, we would expect that as the packing factor increases, the number of synonym chains will increase and that the lengths of the chains will increase. We would like to compare these growth rates for different randomizing algorithms. The information needed is in Figure 16-9.

Randomizer: Name: Byte Limit:		Size of RAA: No. of RAPS:		Data Base: Block Size:	
Packing Factor	No. of Synonym Chains	Minimum Length	Average Length	Maximum Length	Standard Deviation
10	———	———	———	———	———
20	———	———	———	———	———
30	———	———	———	———	———
40	———	———	———	———	———
50	———	———	———	———	———
60	———	———	———	———	———
70	———	———	———	———	———
80	———	———	———	———	———
90	———	———	———	———	———
100	———	———	———	———	———

Figure 16-9. Display of randomizer characteristics.

Graphic results are also desirable, and these are especially useful with conversational online capabilities. For any of the columns in Figure 16-9, such as "No. of Synonym Chains" or "Average Length," the results of various randomizing algorithms can be compared on a single graph (or screen), as in Figure 16-10, and variations can be noted as the randomizing parameters are varied.

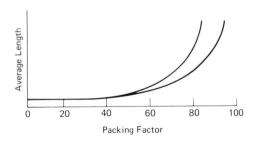

Figure 16-10. Comparative display of randomizer characteristics.

Finally, a presentation like that shown in Figure 16-11 is effective for visualizing the distribution patterns of the root segments. A visual comparison of the root distribution for each of the randomizing techniques can reveal which technique produces the most favorable pattern for the expected key value distributions. Figure 16-11a is a binary presentation showing which root anchor points are empty and which are occupied by randomized root keys. The root anchor point addresses are indicated vertically by hundreds and horizontally by units. Figure 16-11b is a more informative presentation than Figure 16-11a and shows, for entries greater than one, the length of synonym chains and where they begin.

```
                             Units
                0   1   2   3   4   5   6   7   8   9   9
                0   0   0   0   0   0   0   0   0   0   9
           1    1   1   1 1 1   1   1   1         1 1         1
           2    1       1   1111     111     1 1       1   1   1
           3    1   1   1   1   1   111 1     111 11   1   1   1
Hundreds   4   111         111 1  1 1  1          1111     1  111
           5   11    1         1     1     111     11       1  1 1
           *
           *
           *
```

Figure 16-11a

```
                             Units
                0   1   2   3   4   5   6   7   8   9   9
                0   0   0   0   0   0   0   0   0   0   9
           1    1   1   1 2 1   1   1   1         1 3         1
           2    1       1   1231     131     1 2       1   1   1
           3    4   1   1   1   1   111 1     111 11   1   1   1
Hundreds   4   111         112 2 31   1          1421     1  311
           5   21    1         1     5     442   11       1  1 1
           *
           *
           *
```

Figure 16-11b

Figure 16-11. Displays of home address distribution patterns

Watching either of these distribution patterns change as the packing factor increases, which can be done on an interactive display, can be an informative exercise.

17. Space Calculations

PHILOSOPHY

At this point in the design cycle we are ready to begin evaluating the physical design choices made in the previous chapter and the logical design choices made earlier. Physical design consists of choices among such things as access methods, indexing, block sizes, data set groups, root anchor points, root-addressable area size, randomizing algorithms, and pointer options. Computer-assisted techniques can greatly aid in the task of finding the right combination of values for these parameters to produce efficient performance and economic space utilization.

Analytic space and time calculations are a means of evaluating logical and physical designs on a somewhat gross scale. The modeling techniques to be discussed in Chapter 19 can provide a more detailed evaluation. Both approaches have their place. With the space and time calculations, the choices can quickly be narrowed to a few, say three or four, candidate designs. Then, by modeling the application programs and using actual DL/I calls and I/O, the best of these remaining candidates can be determined.

INPUT FOR SPACE CALCULATIONS

The required input for space calculations consists simply of the information normally contained in the DBD source statements plus certain additional information that can be inserted into the DBD source statements. Time calculations also require the patterns of DL/I calls that are to be invoked.

From the DBD statements, we use the names and lengths of the segments, their relationships to each other, and their average frequencies of occurrence. The frequencies specified in the DBD are simply the single-valued average number of segments per parent, or per data base in the case of the root. More precise space calculations are made possible by augmenting the DBD information with the expected frequency distributions of the segments. The more commonly used distributions are the Poisson, Uniform, Bernoulli, and the 80-20 Rule, although others are also occasionally specified. When going to this degree of sophistication, provision should also be made for the designer to submit any nonstandard distributions that apply. Sometimes the presence or absence of a segment depends on the presence or absence of some other "controlling" segment. Provision should also be made for this type of situation. For simplicity in the examples to be presented, we will deal simply with average numbers of segment occurrences and not with individual segment distributions.

SUBTREES

The notion of subtrees is very useful when calculating data base size estimates. A subtree of segment type X is defined to be one occurrence of segment type X plus the collection of all segments that are hierarchically subservient to that occurrence of segment X. Each segment occurrence that has children defines a proper subtree. Each segment occurrence without children defines a primitive subtree consisting of only that segment occurrence.

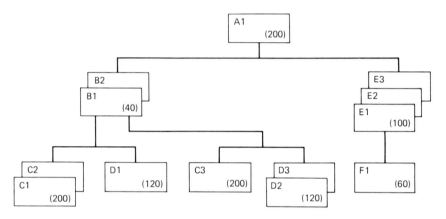

Figure 17-1. Example of a specific data base record (1).

In Figure 17-1, occurrences of segment types C, D, and F are primative subtrees because they are the low-level segments on their respective paths. The subtree of B1 is composed of segments B1, C1, C2, and D1. The subtree of B2 comprises segments B2, C3, D2, and D3. E1 and E3 are also primitive subtrees. E2's subtree contains E2 and F1. A1's subtree is A1 plus the collection of all the B, C, D, E, and F type segments subservient to A1. For purposes of calculation, it is convenient to think of a subtree as containing a defining segment occurrence plus the subtrees of all immediate dependents of that segment.

CALCULATING SUBTREE LENGTH

The length of the subtree of a segment is merely the length of the segment itself plus the lengths of all its dependent subtrees. But it is impractical, and usually impossible, to determine the exact number of occurrences of the dependent segment types because the number of occurrences varies from record to record, and it also varies dynamically within a record as inserts and deletions are made during processing. Hence, we calculate a segment's subtree size on the basis of an "average" record by estimating the expected number of occurrences of each subservient segment type.

For space estimates, maximum values are sometimes used rather than average values for segment occurrence expectations. However, more realistic space estimates can be obtained by using statistical expectations, when known, for the number of segment occurrences. In the examples below, F represents the average expected frequency of occurrence of each segment type and L represents the segment length.

A DL/I segment contains both a prefix and data; the prefix length depends on the storage organization (HD or HS) and, if HD, on the pointer options. For simplicity in the examples to follow, we will assume overall segment lengths without giving detailed consideration to prefix sizes.

Subtree Sizes Without Data Set Groups

For the initial example, assume the data set is not divided into groups. Assume 10,000 occurrences of the root segment A, and assume an "average" data base record, as depicted in Figure 17-2, with segment frequencies (F) and lengths (L) as indicated. (The record depicted in Figure 17-1 can be considered an actual record from this data base.)

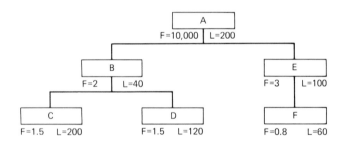

Figure 17-2. Example of an average data base record.

The length calculations proceed in the following manner where $L(X)$, $S(X)$, and $F(X)$ represent the length of segment type X, the length of the subtree of segment type X, and the frequency of occurrence of segment type X, respectively.

$$S(F) = L(F) = 60 \text{ bytes}$$
$$S(E) = L(E) + S(F) \times F(F)$$
$$= 100 + 60 \times 0.8 = 148 \text{ bytes}$$

$$S(D) = L(D) = 120 \text{ bytes}$$
$$S(C) = L(C) = 200 \text{ bytes}$$
$$S(B) = L(B) + S(C) \times F(C) + S(D) \times F(D)$$
$$= 40 + 200 \times 1.5 + 120 \times 1.5 = 520 \text{ bytes}$$
$$S(A) = L(A) + S(B) \times F(B) + S(E) \times F(E)$$
$$= 200 + 520 \times 2 + 148 \times 3 = 1{,}684 \text{ bytes}$$

Thus the expected length of a record (each occurrence of the root, A, and all its dependents) is 1684 bytes. The total number of bytes *(TB)* for the data base is:

$$TB = S(A) \times F(A)$$
$$= 1{,}684 \times 10{,}000$$
$$= 16{,}840{,}000 \text{ bytes.}$$

Subtree Sizes With Data Set Groups

If the data base is divided into groups, the space calculations must be performed on a "per group" basis, as each group will be defined to the system as a separate data set having its own storage characteristics. Using the grouping determined in Chapter 16 for the HD organization, the following example of Figure 17-3 will illustrate space calculations for data set groups. The numbers inside the segments indicate the group to which that segment type was assigned.

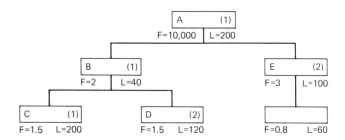

Figure 17-3. Example of a specific data base record (2).

The space calculations for each group are performed as follows:

Group 1

$$S(C1) = L(C1) = 200 \text{ bytes}$$
$$S(B1) = L(B1) + S(C1) \times F(C)$$
$$= 40 + 200 \times 1.5 = 340 \text{ bytes}$$

$$S(A1) = L(A1) + S(B1) \times F(B)$$
$$= 200 + 340 \times 2 = 880 \text{ bytes}$$
$$TB(1) = S(A1) \times F(A)$$
$$= 880 \times 10,000 = 8,800,000 \text{ bytes}$$

Group 2

$$S(E2) = L(E2) = 100 \text{ bytes}$$
$$S(D2) = L(D2) = 120 \text{ bytes}$$
$$S(B2) = S(D2) \times F(D)$$
$$= 120 \times 1.5 = 180 \text{ bytes}$$
$$S(A2) = S(B2) \times E(B) + S(E2) \times F(E)$$
$$= 180 \times 2 + 100 \times 3 = 660 \text{ bytes}$$
$$TB(2) = S(A2) \times F(A)$$
$$= 660 \times 10,000 = 6,600,000 \text{ bytes}$$

Group 3

$$S(F3) = L(F3) = 60 \text{ bytes}$$
$$S(E3) = L(F3) \times F(F)$$
$$= 60 \times 0.8 = 48 \text{ bytes}$$
$$S(A3) = S(F3) \times F(E)$$
$$= 48 \times 3 = 144 \text{ bytes}$$
$$TB(3) = S(A3) \times F(A)$$
$$= 144 \times 10,000 = 1,440,000 \text{ bytes}$$

The total number of bytes for the data base is the sum of the sizes of the data set groups.

$$TB = TB(1) + TB(2) + TB(3)$$
$$= 8,800,000 + 6,600,000 + 1,440,000$$
$$= 16,840,000 \text{ bytes}$$

REQUIRED DISK SPACE

Although we have calculated the number of segment bytes in each data base (or data set group), the required storage space necessitates additional considerations leading to even larger numbers. To calculate required storage space, we have to consider the type of storage device, track capacity, logical record length, block size (or control interval size), and desired free space.

To understand the calculations that follow, the reader must recall from Chapter 15 that blocks are normally not completely filled with segment bytes. When loading segments into a block, there is usually some waste space resulting from not having enough room to accomodate the next segment. In addition, and depending on the access method used, each block contains a number of overhead fields such as free space elements (FSEs), root anchor points (RAPs), control interval definition fields (CIDFs), record definition fields (RDFs), and overflow pointers. Thus, required storage space actually depends on effective storage space for the data bytes plus additional space dictated by waste and by overhead. Although the diagram of Figure 17-4 shows the overhead grouped at the beginning of the block, some of it is actually distributed throughout the block.

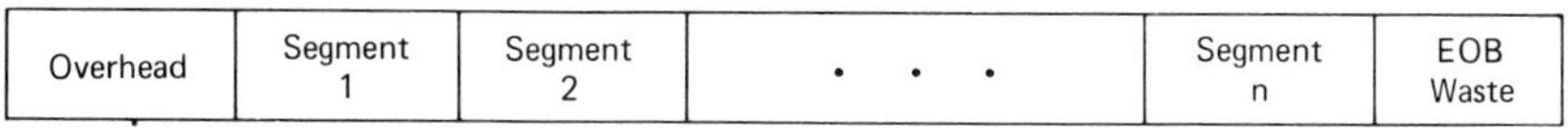

Overhead	Segment 1	Segment 2	. . .	Segment n	EOB Waste

Figure 17-4. Concept of effective storage space.

In addition to overhead and end of block waste, free space can be provided in HD organizations to allow for data base expansion as new segments are inserted during actual processing. To provide for expansion, the designer can designate the percentage of space within a block to be left empty when the data base is loaded (or reorganized). He can also designate a number of blocks distributed through the data base to be left entirely empty. Thus, the number of bytes required for storing a data base depends not only on the number of segment bytes to be stored, but also on the overhead fields, the end of block waste, the free space within the blocks, and the number of initially empty blocks. The basic calculations for estimating storage space follow.

Basic Calculations

The total number of storage bytes required for the data base is given by:

$$TSS = NB \times BLKSZ$$

where:

TSS = Total Storage Space in Bytes
NB = Number of Required Blocks (to be calculated)
$BLKSZ$ = Block Size in Bytes

The number of required blocks (NB) is given by:

$$NB = \frac{TB}{EFFBLK \times (1 - FSB)} \Big] \text{ (next highest integer)}$$

where:

$$TB = \text{Total Number of Segment Bytes}$$
$$EFFBLK = \text{Effective Block Size in Bytes}$$
$$FSB = \text{Percentage of Blocks to Remain Empty}$$

The *EFFBLK* is the number of segment bytes to be stored in a block. This is given by:

$$EFFBLK = (BLKSZ) \times (1 - FSW) - OVHD - EOBW$$

where:

$$FSW = \text{Percentage of Free Space Within the Blocks}$$
$$OVHD = \text{Block Overhead (RAP, FSE, CIDF, RDF, etc.) in Bytes (to be estimated)}$$
$$EOBW = \text{End of Block Waste in Bytes (to be calculated)}$$

The calculation of *EOBW*, which can be complex, is explained in the next subsection.

End of Block Waste

Consider the following probability distribution for segment lengths in the data base:

Length in Bytes	Probability of Occurrence
1	p_1
2	p_2
3	p_3
*	*
*	*
*	*
n	p_n

The expected segment length ($E(L)$) is given by:

$$E(L) = 1 \times p_1 + 2 \times p_2 + 3 \times p_3 + \ldots + n \times p_n$$

The longer the segment, the larger the probability of its being the one that doesn't fit in the remaining space in the block. The probability distribution for the length of the segment that doesn't fit is given by:

Length in Bytes	Probability of Occurrence
1	$p_1 \times 1/E(L)$
2	$p_2 \times 2/E(L)$
3	$p_3 \times 3/E(L)$
*	*
*	*
*	*
n	$p_n \times n/E(L)$

The expected length of this nonfitting segment $(E(L))$ is:

$$E(L') = 1^2 \times p_1/E(L) + 2^2 \times p_2/E(L) + \ldots + n^2 \times p_n/E(L)$$
$$= \frac{1}{E(L)} (1^2 \times p_1 + 2^2 \times p_2 + 3^2 \times p_3 + \ldots + n^2 \times p_n)$$
$$= \frac{E(L^2)}{E(L)}$$

For a nonfitting segment length of L', the probability distrbution for the EOBW space is:

Size of Waste	Probability of Occurrence
1	$1/L'$
2	$1/L'$
3	$1/L'$
*	*
*	*
*	*
$L' - 1$	$1/L'$

Thus, for a given nonfitting segment size, L', the expected amount of end of block waste $(E(W/L'))$ is:

$$E(W|L') = 1 \times 1/L' + 2 \times 1/L' + 3 \times 1/L' + \ldots$$
$$+ (L' - 1) \times 1/L'$$
$$= (1/L') \times (1 + 2 + 3 + \ldots + L' - 1)$$
$$= \frac{1}{L'} \times \frac{(L' - 1)(L')}{2}$$
$$= \frac{L' - 1}{2}$$

The resulting expected waste ($E(W)$), considering all segment sizes, is:

$$E(W) = E(E(W|L')) = \frac{E(L')}{2} - \frac{1}{2}$$
$$= \frac{E(L^2)}{2 \times E(L)} - \frac{1}{2}$$

Thus:

$$EOBW = E(W) = \frac{E(L^2)}{2 \times E(L)} - \frac{1}{2}$$

Example

We will use the logical model example of Figure 17-2 to illustrate the space calculations, and for simplicity, we will assume it is not to be divided into secondary data set groups. If the data base is divided into groups, separate space calculations must be performed for each data set group.

We have already calculated that the TB is 16,840,000. Assume a $BLKSZ$ of 2,048 bytes. To allow for expansion, assume an FSW of 10% and an FSB of 20%. The calculations to follow can apply to HISAM logical records as well as to HD blocks simply by setting FSW and FSB to zero.

We first calculate the estimated $EOBW$. This is done on the basis of a single root segment and its average subtree. The probability distributions of segment length, L, and its square, L^2, are given by the following.

Segment	L	L^2	Probability
A	200	40,000	1/14.4
B	40	1,600	2/14.4
C	200	40,000	3/14.4
D	120	14,400	3/14.4
E	100	10,000	3/14.4
F	60	3,600	2.4/14.4

where the probabilities are obtained by counting the total number of occurrences of each segment type in the average data base record.

The expected segment length is:

$$\begin{aligned} E(L) &= 200 \times 1/14.4 + 40 \times 2/14.4 + 200 \times 3/14.4 + 120 \times 3/14.4 \\ &\quad + 100 \times 3/14.4 + 60 \times 2.4/14.4 \\ &= (1/14.4) \times (200 \times 1 + 40 \times 2 + 200 \times 3 + 120 \times 3 \\ &\quad + 60 \times 2.4) \\ &= (1/14.4) \times 1,684. \end{aligned}$$

Similarly, the expectation of the squared segment length, L^2 is:

$$\begin{aligned}
E(L)^2 &= 40{,}000 \times 1/14.4 + 1{,}6000 \times 2/14.4 + 40{,}000 \times 3/14.4 \\
&\quad + 14{,}000 \times 3/14.4 + 10{,}000 \times 3/14.4 + 3{,}600 \times 2.4/14.4 \\
&= (1/14.4) \times (40{,}000 \times 1 + 1{,}6000 \times 2 + 40{,}000 \times 3 + 14{,}400 \times 3 \\
&\quad + 10{,}000 \times 3 + 3{,}600 \times 2.4) \\
&= (1/14.4) \times 245{,}040.
\end{aligned}$$

Finally, the *EOBW* is given by:

$$\begin{aligned}
EOBW &= \frac{E(L^2)}{2 \times E(L)} - \frac{1}{2} \\
&= \frac{(1/14.4) \times 245{,}040}{2 \times (1/14.4) \times 1{,}684} - \frac{1}{2} \\
&= \frac{245{,}040}{3{,}368} - \frac{1}{2} \\
&= 72.3 \text{ bytes}
\end{aligned}$$

The *EFFBLK* can now be calculated. For this example, we will arbitrarily assume a 31 byte overhead. The actual amount of overhead depends on the DL/I access method, the system access method (VSAM or ISAM/OSAM), and certain internal factors (*FSEs*, number of overflow pointers, etc.).

$$\begin{aligned}
EFFBLK &= (BLKSZ) \times (1 - FSW) - OVHD - EOBW \\
&= 2{,}048 \times (1 - .10) - 31 - 72.3 \\
&= 1{,}740 \text{ bytes}
\end{aligned}$$

The required number of blocks *(NB)* is:

$$\begin{aligned}
NB &= \left\lceil \frac{TB}{EFFBLK \times (1 - FSB)} \right. \\
&= \left\lceil \frac{16{,}840{,}000}{1{,}740 \times (1 - .20)} \right. \\
&= 12{,}098 \text{ blocks.}
\end{aligned}$$

The *TSS* required for storing the 16,840,000 byte data base is now given by:

$$\begin{aligned}
TSS &= NB \times BLKSZ \\
&= 12{,}084 \times 2{,}048 \\
&= 24{,}748{,}032 \text{ bytes.}
\end{aligned}$$

But if the data base grows to fill all the free space initially set aside, this same storage space can accomodate up to 23,491,296 segment bytes.

18. Time Calculations

In this chapter we will use analytic calculations to estimate I/O timings for a data base design. Bear in mind that the data base itself is still a "paper" design which has not yet been implemented on hardware. The purpose of this exercise is to narrow the number of candidate designs by quickly estimating "ball park" times for I/O operations. More precise modeling (Chapter 19) using actual hardware and IMS software can then be used to select the best of the remaining designs. Experience indicates this to be a feasible and useful approach to performance evaluation.

There is a three-fold purpose for the material to be presented in this chapter. First is to indicate the kind of thinking—the concepts and considerations—that must be addressed in doing a performance evaluation of a data base and its use. Secondly, we wish to show that the performance estimates required for these evaluations can be obtained by analytic calculations and some simulation on a computer with greater accuracy than is normally the case with manual techniques. And thirdly, we wish to show that the analytic calculations, at best, are inherently limited in their accuracy; after they have served their purpose, more precise evaluations of the remaining designs requires modeling in a real system environment.

We begin by deriving the probabilities of a physical I/O for the three basic types of I/O operations in IMS. We will then indicate how to combine these probabilities with DL/I call patterns, IMS path lengths, and hardware characteristics to produce a timing analysis of the data base as it will be used.

PROBABILITIES OF I/O

Many simplifying assumptions are made in the derivations to follow. All data base records can be different and, with insert and delete activity, they can vary dynamically with processing. Therefore, the I/O probabilities are derived on the basis of the "average" data base record. The larger the data base and the larger the number of I/Os, the more reliable we can expect these estimates to be. Some additional simplifying assumptions will be explained as they are encountered in the ensuing derivations.

More precise estimates of I/O probabilities can be obtained by considering the unique probability distributions of each segment type, by working with variances as well as with means, and by considering the special characteristics of the access method being used. Dechow and Lundberg have provided criteria and equations for these more precise estimates in Reference D.2.

Types of I/O Probabilities

There are three basic types of I/O operations in IMS (or DOS-DL/I), although there are internal variations in the way these operations are performed depending on the access method. The three operations (Figure 18-1) are: (1) parent to first child, (2) segment to next twin, and (3) segment to parent (for logical data bases).

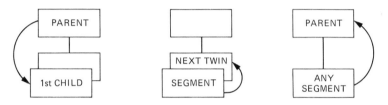

Figure 18-1. Basic types of I/O probabilities.

Therefore, we will calculate the following I/O probabilities:

PCIO—Parent-to-Child I/O

The probability of a physical I/O operation in going from a parent segment to the first occurrence of one of its child segment types.

PTIO—Physical Twin I/O

The probability of a physical I/O operation in going from a segment to its next twin.

PPIO—Physical Parent I/O

The probability of a physical I/O operation in going from a segment to its physical parent.

The calculated probabilities can be presented in tabular form for each segment type as follows:

SEGMENT	PCIO	PTIO	PPIO
Root	X.XX	X.XX	X.XX
Dep1	X.XX	X.XX	X.XX
Dep2	X.XX	X.XX	X.XX
*			
*			
*			

Figure 18-2. Display format for estimated I/O probabilities.

In Figure 18-2, PCIO represents the I/O probabiliy in going *to* the named segment from its parent; PTIO and PPIO represent the I/O probabilities in going *from* the named segment to the next twin and to the physical parent, respectively.

Computer assistance in an interactive encironment can be quite helpful in studying these I/O probabilities and how they will vary with changes in block sizes, free space, access method, and so forth.

The Concept of Distance

Since a block is the unit of information transferred across the I/O channels, the probability that a DL/I call will cause an I/O operation is really the probability that the target segment is in a different block than the source segment at which we are currently positioned. Thus the determination of the I/O probabilities is based on further analysis of the segment lengths and storage patterns developed in the previous chapter. (We are ignoring the possibility that the block containing the target segment is already in the buffer pool from a previous I/O operation.)

The situation is depicted pictorially in Figure 18-3 in which Seg. X represents the source segment where we are currently positioned, and Seg. Y represents the target segment we wish to access.

$L(X)$ and $L(Y)$ represent the segment lengths in bytes. *EFFBLK* is the

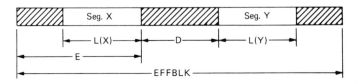

Figure 18-3. Concept of distance between segments.

effective block size in bytes. E is the distance, in bytes, from the beginning of the block to the end of Seg. X, and D is the distance, in bytes, between the end of Seg. X and the beginning of Seg. Y. The inclusive distance between X and Y is denoted by $D(X \rightarrow Y)$ where:

$$D(X \rightarrow Y) = L(X) + D + L(Y)$$

This concept of distance between segments is based on the hierarchical ordering of the segments within a data base record. To a large extent, this hierarchical ordering is present in a newly loaded data base regardless of the DL/I access method. Thus the validity of these analytic time estimates is based on the assumption of a newly loaded data base or at least a data base that still approximates its newly loaded or reorganized storage patterns.

The general pattern of each of the I/O probabilities with respect to effective block size is depicted in Figure 18-4.

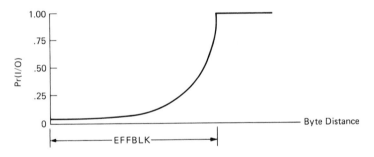

Figure 18-4. Typical I/O probability distribution.

The flat portion of the curve says that if the inclusive distance from the source segment to the target segment exceeds the *EFFBLK*, then there is certainty of a physical I/O operation.

I/O Probability Calculations

Statistically, the segment distribution problem is analogous to the single server queueing model in which the segments represent customers arriving for service, and the segment lengths represent interarrival times. Assuming that Seg. X can occur anywhere in the block, E is a random variable with a uniform distribution. Assuming also that the lengths of the segments, as they are encountered in physical storage, are described by the Poisson distribution, and assuming that the target segment, Seg. Y, is the nth segment following Seg. X, the distance, D, between Seg. X and Seg. Y is a random variable with a Gamma distribution.

The probability of I/O ($Pr(I/O)$) is the probability that Seg. X and Seg. Y are not in the same block. This is given by:

$$Pr(I/O) = Pr(E + D + L(Y) > B)$$

where B is the actual block size.

Without derivation, it is asserted that this probability calculation can reasonably be approximated by:

$$Pr(I/O) = Min \left(1, \frac{D(X \rightarrow Y)}{EFFBLK} \right) \qquad (1)$$

where $EFFBLK$ is the effective block size derived in Chapter 17. Equation (1) applies directly to the calculation of I/O probabilities for PCIO and PTIO types. The calculations for PPIO however, are more complex.

The calculation of inclusive distance from a segment (Y) back to its physical parent (X) is complex because we don't know at which occurrence of Y we are positioned. Assuming we are equally likely to be positioned at any occurrence of Y within X, we should calculate the I/O probability from each of the segment Y occurrences back to the parent X, and then average these probabilities. To do this, we must have some idea of the number of occurrences of Y within X; therefore, if the probability distribution of the occurrences of Y is known, this approach is feasible.

The byte distance (D) from the first occurrence of the current segment type, Y1, back to its parent, X, is given by:

$$D(Y1 \rightarrow X) = D(X \rightarrow Y1)$$

The byte distance from the second occurrence, Y2, back to X is given by:

$$D(Y2 \rightarrow X) = D(X \rightarrow Y1) + S(Y)$$

In general, the byte distance from the nth occurrence, Yn, back to X is given by:

$$D(Yn \rightarrow X) = D(X \rightarrow Y1) + (n - 1) \times S(Y)$$

Applying Equation (1) to the distance from each occurrence of Y back to its parent, X, and averaging, we have:

$$PPIO = PPIO(Y1 \rightarrow X) \times p_1 + PPIO(Y2 \rightarrow X) \times p_2 + \ldots$$

$$+ PPIO(Yn \rightarrow X) \times p_n$$

$$= Min \left(1, \frac{D(Y1 \rightarrow X)}{EFFBLK} \right) \times p_1 + Min \left(1, \frac{D(Y2 \rightarrow X)}{EFFBLK} \right)$$

$$\times p_2 + \ldots + Min \left(1, \frac{D(Yn \rightarrow X)}{EFFBLK} \right) \times p_n \tag{2}$$

where $p_1, p_2, \ldots, p_n$ are the probabilities of the existence of the occurrences of segment Y within X obtained from the probability distribution of Y's occurrences.

On the other hand, when working with an "average" data base record, we must make some simplifying assumptions because we do not know how many actual occurrences of Y may exist for a given parent X. A number of alternatives exist for simplification. The one most commonly used is to calculate the average distance from the Y occurrences back to X and then apply Equation (1) directly to this average distance. Using this approach, the following equations provide the average distance from Y to X.

$$D(Y \rightarrow X) = \frac{1}{F(Y)} \times \begin{bmatrix} D(X \rightarrow Y) & + \\ D(X \rightarrow Y) + S(Y) & + \\ D(X \rightarrow Y) + 2 \times S(Y) & + \\ . \\ . \\ . \\ D(X \rightarrow Y) + (F(Y) - 1) \times S(Y) \end{bmatrix}$$

$$= \frac{1}{F(Y)} \times \left[F(Y) \times D(X \rightarrow Y) + \frac{(F(Y) - 1) \times (F(Y))}{2} \right.$$

$$\left. \times S(Y) \right] \tag{3}$$

$$= D(X \rightarrow Y) + \frac{(F(Y) - 1) \times S(Y)}{2}$$

Equation (1) can now be applied to this average distance to obtain a "ball park" I/O probability from a segment to its physical parent.

Special Considerations

These I/O probabilities apply to searching for a desired segment (establishing position) in a DL/I data base. Thus, they apply primarily to the GET and GET NEXT family of DL/I calls, but also to establishing position for INSERTs (HISAM) and DELETEs. To be complete, searching for indexes and for bit maps should also be considered with INSERTs and DELETEs. For REPLACE operations, the search probabilities are not used because the blocks to be updated have already been located and retrieved. For all updates we are concerned with the probability of physical I/Os for output, which in normal operation, is the probability that a block has received one or more updates.

So far we have considered a dichotomous set of probabilities—the probability of one or more I/Os or of no I/Os. For HD organizations this is sufficient, but with HS organizations, we might want probability distributions of the exact number of I/Os.

In the HD organization, because of the direct pointers, accesses from one segment to another involve at most one I/O because the pointers take us directly to the block containing the desired segment. The main time–consuming factor in processing HD data bases is in following long twin chains from twin to twin (and from block to block) until finding the desired segment.

In HISAM, which is less frequently used than the HD organizations, the sequential ordering of the search might require more than one physical I/O for a DL/I call. This can be the case when processing twins and skipping their dependents. It is here that probability distrbutions for the exact number of I/Os might be desirable. On the other hand, the choice of HISAM is based largely on the concept of processing segments in their hierarchical sequence with a minimum of skipping of intermediate segments. This is to say that with HISAM we will not normally be processing long twin chains unless they are at the lowest level of their respective hierarchical paths.

The foregoing derivations and the example to follow are for probabilities of one or more I/Os per DL/I call. These results are valid (except as noted below), for HD organizations and for HISAM as it is normally used.

Finally, the reader is cautioned that there are certain situations to which these probabilities do not apply. They do not apply to HDAM data bases when the source and target segments are both in the root-addressable area because, by definition, both segments are in the same block. But this consideration is frequently ignored because as the data base is updated and then reorganized, the segment content of the root-addressable area changes. These probabilities also do not apply when the source and target segments are not in the same data set group and also when the I/O traverses a logical relationship to another data base. In both cases, there is certainty of an I/O.

I/O Calculation Example

To illustrate the calculation of these I/O probabilities, we use the "average" data base record depicted in Figure 17-2. This "average" record has segments in the hierarchical order shown in Figure 18-5. The listed frequencies indicate those segments having, on the average, a fractional number of occurrences. Based on this segment pattern, average byte distances for the three basic I/O types are calculated below.

A	B	C	D	B	C	D	E	F	E	F	E	F
$L=$ 200	40	200	120	40	200	120	100	60	100	60	100	60
$F=$		1.5	1.5		1.5	1.5		.8		.8		.8

Figure 18-5. Layout of average data base record.

In this example we assume the HDAM access method with direct accesses to the root segments. Therefore, the calculations to be illustrated pertain only to processing the data segments. If an indexed access method were being used, the calculations should also include time estimates for searching the index.

Parent to First Occurrence of Each Child

The inclusive distance from a parent segment (X) to the first occurrence of one of its child segment types (Y) is the length of X plus the length of Y plus the distance (D) between X and Y. The distance is estimated by the product of the frequency and the subtree size (S) of each of X's siblings that are to the left of Y. The subtree sizes were calculated in Chapter 17. From Figure 18-5 we calculate the following parent-child distances.

$$D(A \rightarrow B) = L(A) + L(B)$$
$$= 200 + 40$$
$$= 240 \text{ bytes}$$
$$D(A \rightarrow E) = L(A) + F(B) \times S(B) + L(E)$$
$$= 200 + 2 \times 520 + 100$$
$$= 1,340 \text{ bytes}$$
$$D(B \rightarrow C) = L(B) + L(C)$$
$$= 40 + 200$$
$$= 240 \text{ bytes}$$
$$D(B \rightarrow D) = L(B) + F(C) \times S(C) + L(D)$$
$$= 40 + 1.5 \times 200 + 120$$
$$= 460 \text{ bytes}$$

$$D(E \rightarrow F) = L(E) + L(F)$$
$$= 100 + 60$$
$$= 160 \text{ bytes}$$

Applying the equation

$$Pr(I/O) = MIN \left(1, \frac{D(X \rightarrow Y)}{EFFBLK} \right) \quad (1)$$

to the distances computed above, we obtain the PCIO probabilities shown in the table of Figure 18-6. In so doing, we are using an *EFFBLK* of 1,740 bytes as calculated in Chapter 17 from a 2,048 byte block. Note in the table that the PCIO for segment A is arbitrarily set to one because there is always an I/O in accessing the root segment. And again, the assumption is that all segments are in the same physical data base and in the same data set group.

Segment	PCIO	PTIO	PPIO
A	1.00		
B	.14		
C	.14		
D	.26		
E	.71		
F	.09		

Figure 18-6. Estimated PCIO probabilities.

Segment to Next Twin

The inclusive distance from a segment occurrence (X) to its next twin (X′) is simply the subtree of X plus the length of X. From the average record in Figure 18-5, we have the following.

$$D(A \rightarrow A') = S(A) + L(A)$$
$$= 1,684 + 200$$
$$= 1,884 \text{ bytes}$$
$$D(B \rightarrow B') = S(B) + L(B)$$
$$= 520 + 40$$
$$= 560 \text{ bytes}$$

$$D(C \rightarrow C') = S(C) + L(C)$$
$$= 200 + 200$$
$$= 400 \text{ bytes}$$
$$D(D \rightarrow D') = S(D) + L(D)$$
$$= 120 + 120$$
$$= 240 \text{ bytes}$$
$$D(E \rightarrow E') = S(E) + L(E)$$
$$= 148 + 100$$
$$= 248 \text{ bytes}$$
$$D(F \rightarrow F') = S(F) + L(F)$$
$$= 60 + 60$$
$$= 120 \text{ bytes}$$

Applying Equation (1) to these distances, we obtain the PTIO probabilities shown in Figure 18-7.

Segment	PCIO	PTIO	PPIO
A	1.00	1.00	
B	.14	.32	
C	.14	23	
D	.26	.14	
E	.71	.14	
F	.09	.07	

Figure 18-7. Estimated PCIO and PTIO probabilities.

Segment to Physical Parent

Applying Equation (3) to the example at hand, the following average distances from a segment back to its physical can be calculated.

$$D(B \rightarrow A) = D(A \rightarrow B) + \frac{(F(B) - 1) \times S(B)}{2}$$
$$= 240 + \frac{(2 - 1) \times 520}{2}$$
$$= 500 \text{ bytes}$$
$$D(C \rightarrow B) = D(B \rightarrow C) + \frac{(F(C) - 1) \times S(C)}{2}$$

$$= 240 + \frac{(1.5 - 1) \times 200}{2}$$

$$= 290 \text{ bytes}$$

$$D(D \rightarrow B) = D(B \rightarrow D) + \frac{(F(D) - 1) \times S(D)}{2}$$

$$= 460 + \frac{(1.5 - 1) \times 120}{2}$$

$$= 490 \text{ bytes}$$

$$D(E \rightarrow A) = D(A \rightarrow E) + \frac{(F(E) - 1) \times S(E)}{2}$$

$$= 1,340 + \frac{(3 - 1) \times 148}{2}$$

$$= 1,488 \text{ bytes}$$

$$D(F \rightarrow E) = D(E \rightarrow F) + \frac{(F(F) - 1) \times S(F)}{2}$$

$$= 160 + \frac{(.8 - 1) \times 60}{2}$$

$$= 154 \text{ bytes}$$

Now applying Equation (1) to these average distances, the PPIO probabilities shown in Figure 18-8 are obtained.

Segment	PCIO	PTIO	PPIO
A	1.00	1.00	—
B	.14	.32	.29
C	.14	.23	.17
D	.26	.14	.28
E	.71	.14	.86
F	.09	.07	.09

Figure 18-8. Estimated PCIO, PTIO, and PPIO probabilities.

EXPECTED DL/I CALLS AND PHYSICAL I/Os

In a sense, the I/O probabilities we have developed serve to evaluate one data base design against another. But a full measure of the "goodness" of a data base design is not achieved until it is evaluated in the context of the way it will be used. Thus we must now consider DL/I call patterns and times for the expected I/O operations.

Up to this point, we have worked with an average data base record to simplify the analytic calculations. But evaluating data base performance evaluation requires additional information about the physical design and the storage patterns of the data base as well as information about the usage to which it will be put. And much of this information can only be obtained by further assumptions or by simulation or modeling.

In order to present the concepts involved and to illustrate them with an example, several simplifying assumptions will be made in the material to follow. These assumptions are heavily dependent on the characteristics of the access method being used as well as on other physical design parameters; hence, the material to follow is more indicative then definitive. Bear in mind that with computer assistance, these assumptions can be avoided by simulating the rules of the data base management system and its actions on the data base. The only inputs needed are the physical and logical DBDs (possibly with supplemental information) of the data bases to be evaluated and the DL/I call patterns to be performed on these data bases.

Still, we are working with averages and probabilities, and the effects of contention are being ignored. Thus, modeling on a computer (see Chapter 19) is recommended when more precise evaluations are required.

DL/I Call Patterns

Assume a sequence of DL/I calls as illustrated in Figure 18-9. For a given Segment A, we wish to update all occurrences of its E segments. We will do this for 100 different As, expecting to find 80% of the As that we seek. For the 20% that are not found, we will insert one A and one E segment.

From the flow chart of Figure 18-9, we see that there will be 100 GET UNIQUEs (GU) to Segment A. Expecting 20% of these calls to return a "not found" status, we expect to do 20 INSERTs (ISRT) of a new Segment A and 20 ISRTs of a new Segment E. Thus, 80 entries are made to the loop for updating the E segments. Since, on the average, there are 3 E segment occurrences each A, we expect to invoke the GET NEXT WITHIN PARENT (GHNP) and the REPLACE (REPL) 240 times each. We expect 80 of the GHNPs to be PCIO (A to E) and 160 to be PTIO (E to E'). Actually, for each set of traversals of the update loop there is a fourth GHNP, making an additional 80 GHNPs, that returns a "not found" status. This GHNP does not do an I/O, so we do not count it later when estimating physical I/Os. These results are recorded in Figure 18-10.

Having determined the expected number of DL/I calls, we now calculate the expected number of physical I/Os and record these results in Figure 18-11.

Each GU to a root segment is classed as a Parent-Child I/O (PCIO = 1). Therefore, 100 I/Os will be expected for the root segment A.

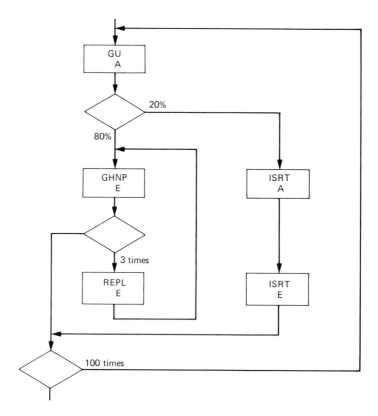

Figure 18-9. Example of a DL/I call pattern.

Segment	GU	GHNP	REPL	ISRT
A	PC 100	———	———	20
B	———	———	———	———
C	———	———	———	———
D	———	———	———	———
E	———	PC 80 PT 160 +80	240	20
F	———	———	———	———

Figure 18-10. Expected number of DL/I calls.

Segment	GU	GHNP	REPL	ISRT
A	100	———	———	38
B	———	———	———	———
C	———	———	———	———
D	———	———	———	———
E	———	80	80	42
F	———	———	———	———

Figure 18-11. Expected number of physical I/Os.

From each A the first access to E will be a Parent-Child (PCIO = .71), and the other two expected accesses to E will be Segment-Twin (PTIO = .14). Thus, the expected number of I/Os for the GHNP to E is 80 × .71 + 160 × .14 = 79.20 or 80.

Since each E that is located will be updated, every retrieved block containing an E will be updated. Hence, the number of physical I/Os for REPLs will equal the same number for GHNPs (i.e. 80).

For the ISRTs, we must make further assumptions. Assume that none of the E segments are in the HDAM root-addressable area. Assume further that there are 5 root segments to an HDAM block, that there are 100 blocks in the root-addressable area, and that there are 200 blocks in the overflow area. These parameters will all have been specified previously when constructing the DBD. Also assume that the blocks receiving inserts remain in core until the end of processing, i.e., IMS synchronous (sync.) point when they all are written out.

Now if a well performing randomizing algorithm is being used, it is equally likely that a root key will randomize to any block in the root-addressable area. Assume there is space for it in the selected block. To calculate the expected number of blocks changed, we first calculate the probability that a given block is not changed, subtract that number from one, and multiply the result by the number of blocks. This yields the following equation:

$$E(\text{Changed Blocks}) = Nb \times \left[1 - \left(\frac{NB - 1}{NB} \right)^{Ni} \right] \qquad (4)$$

where Ni is the number of root segments to be inserted, and Nb is the number of blocks in the root-addressable area. Applying Equation (4) to the root segment inserts, we have:

$$E(\text{Changed Blocks}) = 100 \times \left[1 - \left(\frac{99}{100} \right)^{20} \right]$$
$$= 18 \text{ blocks}$$

Thus, in making 20 root segment inserts, we expect to update 18 blocks. We read the block, update it, and at the end of the process we write it back; therefore, we expect to make $18 \times 2 = 36$ physical I/Os. We don't use the I/O probabilities of Figure 18-8 because, after a physical address is computed by the randomizing routine, reference is made directly to the addressed block after first consulting the bit map. We also may update the bit map which is in core at this time. Allowing two more I/Os, one to read the bit map at the beginning of the process and one to write the bit map at the end of the process, we now estimate a total of 38 physical I/Os for inserting the 20 root segments. To be complete, we should also consider that inserts may cause additional I/Os for updating indexes. With HDAM there is no index, so we finalize the estimate of 38 physical I/Os.

Estimating the number of I/Os for dependent segments is even more elusive. IMS would like to place the E segments into the same block as their root segments, but we have already assumed a byte limit that does not admit the E segments into the root-addressable area blocks. Therefore, on the basis of 20 inserts of E uniformly distributed over 200 overflow area blocks, Equation (4) can be employed again to yield an expectation of 19 blocks changed. Doubling this number gives 38 I/Os to read and to write each of these blocks. But in addition to inserting the segment itself, pointers in parent segments and in surrounding twin segments (and in child segments when present) must be updated when using the HD organization, and the blocks containing these segments may not already be in core. (Or if using HISAM, segments may even have to be moved.) We will assume that, of the 20 inserts of segment E, 20% will require an additional I/O. Thus, the total number of I/Os is 42. Again, updates to an index are not a factor with HDAM, and updates to the bit map need not cause I/Os beyond those accounted for above. These estimates are summarized in Figure 18-11.

At best, the accuracy of estimates such as those used above is quite uncertain. Physical I/O probabilities for inserts (and for deletes) are very difficult to estimate precisely. If they are critical to the evaluation, the modeling processes to be discussed in the next chapter are recommended. But the estimates made thus far are usually adequate for gross evaluations, and the I/O probabilities for the three basic I/O operations are particularly helpful when the searching of long twin chains is a critical factor.

CPU AND PHYSICAL I/O TIMINGS

Having estimated the number of DL/I calls and the number of physical I/Os, we now turn our attention to the questions of CPU times and I/O times for the DL/I calls. Up to this point we have accompanied these performance concepts with a numerical example to illustrate the computations that can be automated and to show the assumptions (or simulation) that are required. But to avoid specifying speeds and capacities for specific hardware and software products, we will merely indicate the timing calculations.

For each type of DL/I call against a segment, CPU time is estimated by multiplying the number of calls, the path length, and the CPU speed. We have:

$$\text{CPU Time} = E(\text{DL/I}) \times PL \times MIPS$$

where

$E(\text{DL/I})$ = Expected Number of DL/I Calls (from Figure 18-10)
   PL = Path Length (average number of instructions for the call)
$MIPS$ = CPU Speed (usually given in instructions per microsecond)

I/O time is obtained by multplying the number of physical I/Os by the sum of the expected seek time, the latency time, and the transfer time.

$$\text{I/O Time} = E(\text{I/O}) \times (ST + LT + XT)$$

where

$E(\text{I/O})$ = Expected Physical I/Os (from Figure 18-11)
   ST = Seek Time (see below)
   LT = Latency Time (time for 1/2 disk revolution)
   XT = Transfer Time (block size divided by device speed)

Seek time is obtained by estimating the number of seeks and the length of each seek. The length is estimated in terms of cylinders traveled and then converted to travel time. These parameters depend on the storage patterns, the device characteristics, and the pattern of DL/I calls from which the arm movement back and forth between segments can be derived. They can roughly be estimated analytically or by simulation for nonshared devices. For shared devices where other activities are competing for the same arm, seek time is best obtained by modeling, although it is often estimated as a single overall average value.

The most important areas of concern in performance evaluations are the portions of the data base involved in the most frequent number of accesses or in the searching of long twin chains or in both. It is by comparing the times or the probabilities for these areas against their counterparts from other design or call pattern variations that we determine which design to accept, even though the rejected design may have been better in some of the less frequent activities. Computer assistance can provide the information, but at the current state of the art it is still best for the human designer to make the judgments.

PATH VERIFICATION

Although this chapter is devoted to timing calculations, an important side product is also available. Having defined the data base design (by DBD) and having specified the DL/I calls to be made, path verification can also be performed.

It is important to determine whether the data required by each application function can, in fact, be accessed. We stated at the conclusion of the logical design phase that the resulting logical design satisfied the functional requirements. Structurally, this is true, but now that we have chosen access methods, pointer options, indexing, and so forth, it is now worthwhile to verify that the rules of the data base management system have not been violated.

Indeed, path verification can be another criterion for eliminating some design variations from further consideration. The output can include a list of DL/I call types for which the desired segment cannot be accessed, and it can serve as a basis for further evaluation.

SUMMARY

For these analytic calculations, necessary simplifying assumptions have been made such as average data base record size and average distance traveled; because of these assumptions, there is a rather wide confidence band about the calculated results. These calculations deal with CPU timings and physical I/O timings, but they give no information about contention aspects such as lockouts, channel and device busy, or increased seek times.

The purpose of this chapter is to suggest a methodology and a technique. More precise results are possible from more precise assumptions and derivations. However, experience indicates that the methods herein illustrated are effective in helping eliminate the more grossly inefficient designs. The remaining design variations can then be thoroughly evaluated under actual operating conditions by the modeling to be discussed in Chapter 19.

19. Application Program Modeling

PHILOSOPHY

We have seen, in Chapter 18, some of the difficulties of estimating values and probabilities of the various parameters that must be considered in data base performance evaluation. More precise evaluations can be obtained by modeling the data base and its application functions in a "live" environment.

Evaluation

To be more precisely evaluated, the design of a data base must be studied in the context of the use to which it will be put. In the sterile environment of analytic evaluation, we can determine on a gross level that one data base design requires less space than another or that one design should perform better than another. But exercising prototype models of the remaining candidate designs and performing I/O calls via the actual data base management system is still the best means of determining the final design. Working with actual DL/I calls, with actual data base records rather than with "average" records, and in an environment where normal contention for data base resources is present, we can chose the most appropriate design with a high degree of confidence.

With regard to DL/I data bases, we propose creating and loading a prototype model of the data base, creating models of the application functions to use the data base, executing these models in an IMS (or DOS-DL/I) system, performing actual DL/I calls against the data base, collecting statistics of times and counts and buffer pool activity at appropriate places, and evaluating the results. We are actually evaluating two different, but related, things: the data base design itself and the application program's use of the data base.

Data Base Design Evaluation

Evaluation of the data base involves an evaluation of both the logical and physical design criteria. It can also be an evaluation of the ability of the data base to grow. For example, the effects of adding or deleting segment occurrences, adding or deleting segment types, modifying segments, and adding new reports and new functions can be evaluated and compared against current requirements. Also, variations in logical relationships, secondary indexing, or searching can be studied. Physical vs. virtual pairing can be evaluated, as can left-to-right ordering of segments or variations of the hierarchical structure. DBD parameters such as pointer options, block (or logical record) sizes, size of root-addressable area, and the number of root anchor points, can similarly be evaluated.

Application Program Call Pattern Evaluation

In many cases of poor data base performance, the culprit has not been the data base itself but the use being made of it. As a simple example, if an application program begins an update sequence with a GHU followed by a REPL and 80% of the time the desired segment is not there but must be inserted, the program is making inefficient use of the structure and existing content of the data base. Further, by collecting activity against certain segments, the call patterns may be able to take advantage of current buffer pool contents to reduce the number of actual physical I/Os. Variations of DL/I call patterns and buffer pool activity can be evaluated for given data base designs.

Types of Modeling

Two types, or levels of sophistication, of modeling are recommended: (1) a preliminary modeling simply for path verification and (2) a detailed modeling for performance analysis.

Path Verification

Even though path verification can be performed by analytic methods, confirmation can be obtained at very little added cost by actual DL/I calls. Each application function's ability to accesss its desired segment types can be reported along with statistics about the number of successful and unsuccessful calls for each segment type, and diagnostics can be provided for call failures.

Detailed Timing Evaluation

Detailed timings of the I/O calls constitute the heart of the modeling concept. Under actual operating conditions, we want to evaluate how much CPU time and elapsed time are required for each type of DL/I call, for groups of calls, and for the entire function. We also want to compare these times with those obtained from variations in the data base design or size or in the application's use of the data base or in the buffer pool activity.

The Evaluation Procedure

We take the position at this time that performance evaluation is a subjective rather than an objective endeavor, and that it is best accomplished by human judgment rather than by automated calculation. There is no one parameter that best characterizes data base performance. Also, to select a specific group of parameters, assign weights to them, and process them to obtain a single measure of performance, is itself a subjective exercise. At the present state of the art, the role of computer assistance is in the modeling. The evaluation still belongs to the human designer. But if the computer assistance has done its job properly, the designer has the kind of information he needs and in the formats he needs.

Therefore, we are recommending the following approach to modeling and evaluation as illustrated in Figure 19-1. The modeling and the reporting of the results are the automated parts of the process. The rest, especially observing and comparing the results, is a human endeavor.

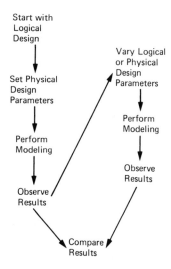

Figure 19-1. Manual comparison of modeling results.

MODELING INPUTS

Two types of input are required for the modeling procedures. A description of the data base(s) to be modeled is required, along with a description of the application program's expected use of the data base(s).

Data Base

As with the analytic space and time estimates, the data base to be modeled can be described by DBD statements with appropriate supplemental information. Primarily, this additional information will consist of segment distributions and key value distributions. We have already discussed the reasons for providing segment distributions. The key value distributions are used to control the probability of finding the desired segment in each data base call. This also means they can be used to guarantee a "hit" for every call. From this type of input, the modeling procedures can automatically allocate and load the model data bases.

Application Program

A working model of the application programs must also be provided. What is needed are programs in load module form that invoke actual DL/I calls against the prototype data base, simulate the logic of the application, and perform measurements from which I/O timings and other performance statistics can be derived. Macro extensions to the common programming languages (COBOL, PL/I, Assembler, etc.) can be provided to simplify the task of preparing such models. Special modeling languages have also been devised for this purpose. One advantage of using the macro extensions is that when the modeling is completed, a running program is in existence in a language familiar to the application programmers. By removing the timing and measurement macros and inserting detailed programming logic, a production application program can be very easily obtained. In addition to the application model, of course, a PSB to describe the model's data base requirements must also be provided.

APPLICATION PROGRAM MODELING CONSIDERATIONS

To provide the desired measurements for performance evaluation, the application program models must contain certain features. The features (or capabilities) of primary interest are suggested below and then illustrated by an example.

Basic Modeling Features

The objective in application program modeling is to prepare a prototype model of an application program that simulates as closely as possible the program's logic and call pattern structure and that collects useful performance information. Six basic types of modeling features are vital to such a prototype.

1. Actual DL/I Calls. The major ingredients of the prototype model are the DL/I call patterns to be executed. The primary purpose is to evaluate the efficiency of these call patterns on the data base that has been designed and to determine areas for tuning or redesign.
2. Time Delays. Time delays are used to approximate the time required for internal application processing between DL/I calls.
3. Branching. Branching is used to simulate the decision-making and looping of the application. Various types of branching can be used. I/O branching can be based on DL/I status codes. Statistical branching (e.g., 80% to A, 20% to B) can simulate random processes. Value branching can be done based on the value of parameters, counts, or other statistics.
4. Program Statistics Collection. Various statistics can be collected. Elapsed and CPU times of individual DL/I calls, of groups of calls, and of entire processes can be collected. Counts of the number of traversals of selected paths can be gathered. Values of interesting parameters from status registers or from calculations can be recorded at specified time intervals or for specified path traversals. Besides being reported, these statistics can be accumulated and compared, and branching within the application model can be done accordingly. The capability should be present for these statistics to be accumulated for the duration of the run or to be reset at specified times.
5. Buffer Pool Statistics Collection. The contents of the VSAM and the ISAM/OSAM buffer pools should also be collected at appropriate intervals in order to study the actual physical I/Os resulting from the DL/I calls.
6. Statistics Printing. Macros should be available to print selected statistics at any point in the model. They can be reported graphically or tabularly, and in detail or as means, deviations, maximums, and so forth. The option of resetting or continuing to accumulate should be part of this feature.

Modeling Example

Figure 19-2 gives a simple example of the use of most of these basic features in application program modeling. This is an example of an update operation in which the application program expects the required segment to be present most of the time. It does a GU and expects to follow with a REPL. Should the GU fail, it inserts a new segment. The entire update process is repeated 100 times. To approximate internal processing times, time delay macros are included before the ISRT (10 macroseconds) and after each update iteration (50 macroseconds). Arrows indicate places to insert additional macros to record the times of each I/O operation, time for the overall process, and counts of traversals of

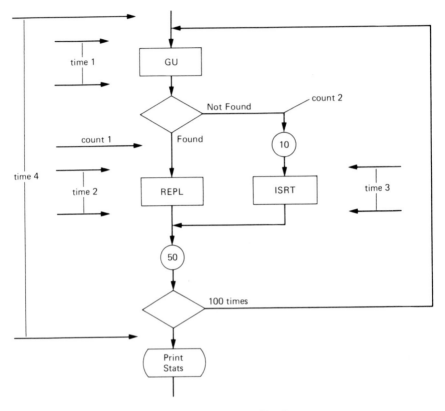

Figure 19-2. Example of a DL/I call pattern.

the REPL and the ISRT paths. Finally, after 100 iterations, selected statistics are printed.

DATA BASE MODELING

The following are the primary considerations required for successful modeling of the data base itself.

Generation and Loading

From the physical and logical DBDs (and their supplemental information), the modeling programs can derive all the needed data base characteristics, and the data bases can be generated, allocated, and loaded automatically. The necessary Job Control Language can be automatically generated to perform these functions, calling on the IMS utility programs as needed.

Segment Content

The segments loaded into these newly created data bases need contain only key field values. (Here we are using "key" loosely to mean all search fields regardless of whether or not they are defined as sequence fields.) Values for nonkey fields need not be included. The purpose of the application model is to evaluate data base performance; therefore, it need not be interested in the segment contents after segments have been located and retrieved.

Segment Key Values

The designer must specify (directly or indirectly) the values of the key fields for the occurrences of each segment type to be accessed. The ability should be present for the designer to supply exact key values or to specify statistical or empirical distributions from which the key values can be automatically generated. For example, the designer may want the key field of one set of segment occurrences to contain values generated randomly in increments of 10 and uniformly distributed between 0 and 80. In response to this specification, the modeling programs might use a random number generator to generate key values of 10, 20, 10, 40, 50, 20, 60, 60, 50, 40, 0, 70, 40, 30, 60, 80, etc., depending on the number of segment occurrences.

While generating these key values, the generating routines can build tables of the generated key values for use in controlling the hit ratios of the DL/I calls to be made against the various segments.

Random and Guaranteed Hits

The capability must exist for specifying random (or statistical) hits for certain segment types. For example, the designer may want to specify that the occurrences of a certain segment type are to be found 80% of the time and not found 20% of the time. Assuming the segment key values are distributed from 0–80 in increments of 10 (as illustrated above), this hit ratio can be accomplished by generating random key values for the Segment Search Argument (SSA) that range from 0–100 in increments of 10. Many other types of hit ratios can also be obtained according to desired statistical patterns or other criteria.

On the other hand, tables of the existing key values can be used to assure guaranteed hits. By randomly or sequentially selecting a key value from the appropriate table and inserting it into the SSA as a search argument, random or sequential accesses can be made with the hit guaranteed. There are at least three places where guaranteed hits are required: (1) locating logical parents, (2) locating targets of secondary indexes, and (3) locating segment occurrences that are otherwise known to exist. For these purposes, tables of concatenated key values can be built by the key generating routines.

Finally, making additional accesses to segment occurrences previously accessed can be accomplished by having the application model save their concatenated keys for future use.

Physical Pairing

If physically paired segments are to be present in the data bases, special care is required in generating their key values to assure that a search key that produces a hit in one will also produce a hit in the other and vice versa. Note that the concatenated keys will *not* be the same. Reference D.4 describes one technique for obtaining key compatibility between the paired segment occurrences.

OUTPUTS

Many types of information can be reported by the application model to show the performance characteristics of the data base as it is being used. The following is a representative sampling of available information that can be useful for performance evaluation. It can be presented in tabular form as detailed lists or as means, distributions, maxima, and minima. Graphic presentations and histograms are especially recommended.

I/O And Program Timings

CPU and elapsed times can be reported by segment type and call type for specified calls and segments. They can also be reported for groups of call sequences, for portions of the application model, and for the entire processing of the function.

DL/I status code summaries can also be presented showing the patterns of successful and unsuccesful calls for the various segment types.

DL/I Call Analyses

DL/I calls to selected segments can be selectively traced by printing the contents of the Program Communication Block (PCB), Input-Output Area (IOA), and the SSA. Both the status code and the key feedback area from the PCB are especially useful in analyzing the call activity.

Counts And Values

Counts and ratios of traversals of selected paths in the model, as well as counts and statistics of the number of executions of selected program instructions, can be reported. Values of registers and results of calculations within the application model can provide information about situations of special interest to the designer. Current values within the run and final values at the end of the run should be available.

Buffer Pool Statistics

VSAM and ISAM/OSAM buffer pool statistics provide especially helpful information for performance evaluation. The number and sizes of the buffer pools are critical in determining how frequently blocks are written out to make room for other I/O and then read back in for further accessing. These statistics can be printed at the end of the run, at checkpoint times, or at other specified intervals.

20. Guidelines for Practical Physical Design

GENERAL

To conclude this treatment of physical design of data bases, a number of practical design guidelines will be presented in summary form. These guidelines, intended to help enhance DL/I data base efficiency with respect to performance and space utilization, are gathered from several sources. Good detailed treatments are found in Dechow and Lundberg (Reference D.2) and in McElreath (Reference A.6)

Although they are by no means all-inclusive, the guidelines presented here represent necessary considerations required of the designer when making and evaluating physical design choices (see Chapters 15 and 16) and in refining the logical design (see Chapter 12). Many of these guidelines have been treated, directly or indirectly, within the preceeding text, although a few are new concepts. While these guidelines are very specific to DL/I data structures and to the IMS data base management system, they should suggest the considerations required for efficient data bases under other data base management systems.

The guidelines are grouped into the categories shown below. An access method category is not included because of the treatment of this topic in Chapter 15.

- Data bases
- Segments
- Physical twin chains
- Pointer options
- Logical relationships
- Secondary indexes

DATA BASES

- Avoid long data base records. Records that fit within a block are ideal in terms of processing time although they may waste storage space.
- Do not create many data bases. This will simplify such things as backup, recovery, reorganization, and control block overhead.
- When data bases are too large to be reorganized in the time available, use key ranges as the criteria for splitting.

- One way to reduce contention from program isolation is to increase the number of roots or to place some of the data into a main storage table.
- Compaction decreases storage requirements and also tends to decrease physical I/Os by combining more segments into a block. The trade-off is CPU time for the compaction routines.

SEGMENTS

- Frequently accessed segments should be kept as close to their root as possible. This decreases the probability of physical I/Os when accessing them because it increases the likelihood of their being in the same block along with the root. They should be structured to the left of less frequently accessed children of the same parent.
- Fewer segment types generally mean less reorganization time.
- If a segment prefix is larger than the data portion, consider combining it with its parent to save space.
- Combining segments to improve performance, or for other reasons, can result in less data independence and increased application program maintenance. In an update environment, it can also result in unwanted redundancies.
- Segments of varying sizes should not be placed into the same data set group if frequent inserts or deletes or both are to be performed. Space is made available for a new segment only if there is enough for the largest segment.
- Consider a variable length segment if a parent has one occurrence of a child segment type that may or may not exist.
- Avoid variable length segments when updates may increase a segment's length. The longer segment may have to be split.

PHYSICAL TWIN CHAINS

- To reduce the time for searching long twin chains, use secondary indexing or break up the twins into groups by structuring "index" segments as parents of each group.
- Use nonkeyed segments to reduce overhead of inserts or deletes or both within twin chains below the root.

POINTER OPTIONS

- Physical child and physical twin pointers are most frequently used.
- Use backward pointers only for long twin chains subject to frequent updates.

- Direct pointers generally imply fewer physical I/Os.
- Symbolic pointers, except with virtual pairing, generally allow independent reorganization of data bases.

LOGICAL RELATIONSHIPS

- For bidirectional virtual pairing, place the real logical child under the parent in the most frequently used path to the child.
- Bidirectional physical pairing is usually most appropriate when accessing intersection data from both parents and when being concerned with the sequential ordering of the destination parent.
- Searching logical twin chains can be costly. The probability of a physical I/O for each twin is high.

SECONDARY INDEXES

- Secondary indexes are suitable for frequent accesses to a small percentage of data base records.
- Avoid indexing to target fields that are subject to frequent updates.
- Sparse indexes tend to reduce the number of physical I/Os to find a target. This may be most efficient if intermediate targets are not wanted or can be accessed sequentially.

PART V
ANCILLARY DESIGN
CONSIDERATIONS

In addition to the design concepts and procedures that have been presented, certain ancillary considerations arise when dealing with the actual practice of data base design. This book will address two topics in this category: desirable interfaces with a data dictionary and a possible interactive data base design language. In both cases the discussion is intended to be suggestive rather than definitive. The real purpose of this section is to stimulate thought on the part of the reader.

21. Desirable Dictionary Interfaces

A dictionary system is normally regarded as a mechanism for storing information about the data in existing data bases. As a repository for descriptions of the data, a dictionary will contain the names of the data elements, verbal descriptions of their meanings, and descriptions of their characteristics (length, format, etc.). Alternate names (synonyms) for the same data are maintained and controlled. Information regarding ownership, security, and update control of the data is also maintained. In addition, dictionaries contain where-used information relating data to programs and to program owners so that those affected can be identified whenever changes to the data base are proposed.

Dictionaries also contain descriptions of the structure of the data base. With regard to IMS, they contain descriptions showing the relationships of data elements to segments, segments to physical data bases, and physical data bases to logical data bases. In this respect, they contain sufficient information for generating physical and logical DBDs.

A dictionary system and an automated data base design process that are closely integrated have much to offer one another. They both deal with descriptions and characteristics of data elements and with the logical structures obtained from these elements and their relationships. A dictionary can provide input to the design process and the design process can store its results into a dictionary.

The desirable interfaces between a dictionary and a data base design process fall into three categories:

- Initial data entry and editing
- Logical model structuring
- Physical model structuring

INITIAL DATA ENTRY AND EDITING

Data Entry

The data requirements information needed by computer-assisted data base design procedures is almost a complete (proper) subset of the information normally stored in current commercial dictionary systems. If the unstructured data elements can be related to their local views, and if the association types between related pairs of unstructured data elements can be included with the relative frequencies of use of these associations, then the dictionary descriptions can serve as input to the automated design procedures. But this information about unstructured data elements is not currently supported in many of the commercially available dictionaries.

Potentially speaking, duplicate entry of essentially the same information into a dictionary system and also into an automated design procedure can and should be avoided. The duplication becomes especially critical when dealing with large data bases of hundreds or even thousands of elements. Entry of information about raw data elements should be made to the dictionary system and an interface should exist to allow the design procedures to access this information in named aggregations (local views).

Editing

Initial data entry is rarely clean—names, usage, and characteristics of the data elements may not yet be standardized across local views. Synonyms, homonyms, and inconsistent characteristics of the same data usually result when data requirements are gathered from different sources. The editing phases of the conceptual design process, and the reports produced therein, can serve as an input filtering function for the dictionary. When the iterative editing phases are completed, obsolete information (e.g., nonstandard names) can be removed from the dictionary. The information remaining permanently is clean and consistent. Synonyms that the designers desire to retain can be maintained in a controlled manner with reasons for their retention being part of their textual descriptions.

LOGICAL MODEL STRUCTURING

Initial Design

The structuring procedure should be able to extract filtered, unstructured data element information in named aggregates (local views) from the dictionary so that the composite model and the derived logical designs can be generated in the normal manner. After diagnostics have been resolved and refinements made, the results of the logical design can be stored into the dictionary.

Storing the Results of a Logical Design

Assuming the unstructured data elements are already described in the dictionary, the relationships defining segments, data bases, logical relations, and secondary indexes would now be stored. Naming the segments and the resulting data bases, while not a part of the design procedures that have been described, can be performed prior to dictionary storage by means of special structure specification commands similar to those suggested in Chapter 12. Final designer decisions regarding which logical relationships and which secondary indexes to implement can be communicated to the design procedures in the same manner. Thus, the dictionary can become a repository for the candidate logical designs awaiting physical design and evaluation.

An additional desirable feature is the ability to delete from the dictionary the descriptions of earlier candidate logical designs.

Adding New Requirements to Existing Designs

When processing new functions or adding new data to an existing data base, the design process should be able to extract from the dictionary a description of the existing design with the filtered, unstructured, data element information for that which is new. A set of constraints must also be imposed on the freedom of the structuring process to restructure the existing design. On one hand, no restructuring should be permitted. The data base already exists and is in productive use, and the designer merely wants to know how well the new requirements are supported by the existing data base and where conflicts may exist. On the other hand, the enterprise may not yet be locked into the existing

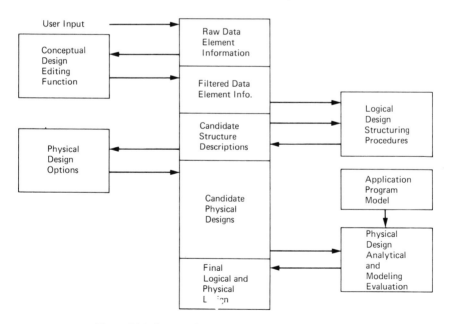

Figure 21-1. Suggested interfaces with a dictionary system.

design, and the designer wants to allow full freedom of restructuring for best overall results.

Additional constraint levels between these two extremes may also be desirable. For example, the designer may want to constrain the hierarchical relationships while allowing segment contents to vary. Variations of segment content may have a relatively minor impact on existing programs since it usually requires only a macro update and a recompilation; changes to the hierarchical relationships can require a redesign of program logic. The designer may also want the ability to constrain certain local views from change while allowing other local views to be modified.

PHYSICAL MODEL STRUCTURING

The computer assistance described in Chapter 14 for determining the physical design options (device type, access method, block sizes, data set groups, pointer options, etc.) can be designed to place these choices appropriately into the dictionary in order that test DBDs can be generated for the physical design procedures. These test DBDs from the dictionary, along with source code for the application model, will serve as input for the physical design procedures. After the designer has selected the final logical and physical design, he should be able to communicate that decision to the dictionary and delete (optionally) the previous candidate designs. This information flow is depicted in Figure 21-1.

22. An Interactive Data Base Design Language

OVERVIEW

The Data Base Design Language is a suggested interactive terminal language for the two basic functions of entering the data requirements and controlling the processes of editing and logical design. The proposed language is intended merely to be illustrative of the kind of interactive capability that would be convenient for this aspect of data base design. It can be considered as a suggested model, and not as a substitute, for a portion of a good dictionary language.

The language is basically a menu-driven language. A hierarchy of menus and formatted screens are provided, but the experienced user should have the capability of skipping intermediate levels. For illustrative purposes, features and capabilities of the IBM 3270 terminal will be used in describing the language.

The language is designed to:

- Minimize the amount of keying, especially for the high-volume initial data input
- Be simple to learn and understand

Data Entry and Editing

Initial data entry can be voluminous. The screens are designed to simplify and minimize the task, menus are used to request displays of editing results, and simple screens permit initial data entry and editing changes.

Structuring and Refinement

Structuring and refinement is an iterative process for which an interactive capability is desired. Menus permit easy display of structuring diagnostics and of design results. Simple screens permit the designer to specify his constraints and refinements.

Certain special editing features will be helpful. For example, in addition to ordinary local and global editing, the ability to apply changes to selected local views and not to others is useful. Certain functions that can be invoked by a single keystroke (e.g., function key) would also be useful. One example is moving the cursor to a given parent-child relationship on the Parent-Child graph, hitting the function key, and getting a list of local views from which that relationship was derived.

INITIAL MENU

The first menu (Figure 22-1) is used to select the basic functions to be performed. The terminal operator selects the desired function by keying its number after the ⟹ or by keying an "S" before the desired report number.

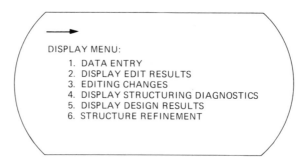

DISPLAY MENU:
    1. DATA ENTRY
    2. DISPLAY EDIT RESULTS
    3. EDITING CHANGES
    4. DISPLAY STRUCTURING DIAGNOSTICS
    5. DISPLAY DESIGN RESULTS
    6. STRUCTURE REFINEMENT

Figure 22-1. Initial menu.

In the screen illustrated in Figure 22-1, the following functions are provided:

- Data entry. Initial entry of data element names and their relationships as specified in the local views.
- Display edit results. Display selected lists prepared by the input editing function. Hard copy may also be requested.
- Editing changes. Correction or alteration or both of local views as indicated by the edit results or by further human analysis.
- Display structuring diagnostics. Display the diagnostic reports prepared by the structuring processes. Hard copy can be requested.
- Display design results. Display the design reports showing the suggested logical design derived by the structuring process. Hard copy can be requested.
- Structure refinement. Specify design constraints and alterations to be imposed on the structuring process.

DATA ENTRY

For data entry, which normally is high-volume input, the screen of Figure 22-2 is presented to the operator.

Data requirements are entered on the basis of a local view. Along with its name, the local view is described by its type (i.e., input, output, or processing),

DATA							ENTRY:			
LOCAL VIEW NAME				TYPE	B/OL		FREQ		PERIOD	

F	T	DATA ELEMENT NAME	A	FREQ	OCCR	T	LENG	F	FACT
.	.	. . . .	.	. . . . .	. . . . .	.	. . . . .	.	. . . . .
.	.	. . . .	.	. . . . .	. . . . .	.	. . . . .	.	. . . . .
.	.	. . . .	.	. . . . .	. . . . .	.	. . . . .	.	. . . . .
.	.	. . . .	.	. . . . .	. . . . .	.	. . . . .	.	. . . . .
.	.	. . . .	.	. . . . .	. . . . .	.	. . . . .	.	. . . . .
.	.	. . . .	.	. . . . .	. . . . .	.	. . . . .	.	. . . . .
.	.	. . . .	.	. . . . .	. . . . .	.	. . . . .	.	. . . . .
.	.	. . . .	.	. . . . .	. . . . .	.	. . . . .	.	. . . . .
.	.	. . . .	.	. . . . .	. . . . .	.	. . . . .	.	. . . . .
.	.	. . . .	.	. . . . .	. . . . .	.	. . . . .	.	. . . . .
.	.	. . . .	.	. . . . .	. . . . .	.	. . . . .	.	. . . . .
.	.	. . . .	.	. . . . .	. . . . .	.	. . . . .	.	. . . . .
.	.	. . . .	.	. . . . .	. . . . .	.	. . . . .	.	. . . . .
.	.	. . . .	.	. . . . .	. . . . .	.	. . . . .	.	. . . . .
.	.	. . . .	.	. . . . .	. . . . .	.	. . . . .	.	. . . . .
.	.	. . . .	.	. . . . .	. . . . .	.	. . . . .	.	. . . . .
.	.	. . . .	.	. . . . .	. . . . .	.	. . . . .	.	. . . . .
.	.	. . . .	.	. . . . .	. . . . .	.	. . . . .	.	. . . . .

Figure 22-2. Data entry screen.

whether it is a batch or online function, and how often it will be processed in a given time period.

For the data element names, the first two columns, "F" and "T", are used to indicate, for a binary relationship, which are the "from" and "to" elements. Their use will be indicated by example below. Following the data element names, the columns have the following meanings:

A	—	Association Type
FREQ	—	Frequency of Use
OCCUR	—	Expected Number of Occurrences
T	—	Data Type (alpha, numeric, etc.)
LENG	—	Data Length
F	—	Format Type (fixed or variable length)
FACT	—	Indicators of which other elements, if any, participate in calculating the value of this element

Two types of information must always be entered: local view name and data element names. All other fields are optional. The following rules apply:

1. Indentation of fields indicates data relationships as depicted on the bubble charts. The "F" and "T" fields can also be used to indicate data relationships.
2. For each association indicated by indentation, the editor will derive and supply the most likely association type based on the indentation patterns. The operator should review these and override these types that are incorrect.
3. Indentation and field-to-field movement can be done by tabbing.
4. Scrolling up and down can be done by function keys.
5. The ENTER key signifies the end of that local view. A new screen is presented for entry of the next local view.

In illustrating the use of the data entry tabular format, the example of a local view given in Figure 22-3 will be used. The data entry is represented by

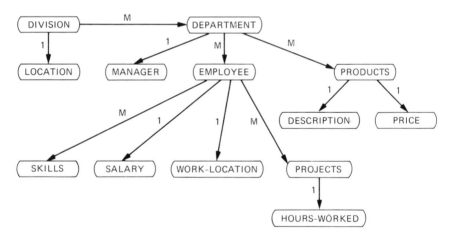

Figure 22-3. Example of a local view.

the screen in Figure 22-4. This screen represents the minimum amount of keying necessary.

Now suppose the operator wants to add an additional element, BIRTH-DATE, to be identified by EMPLOYEE. Adding it at the end of the list with proper indentation will not work because it would be interpreted as being functionally dependent on LOCATION or PRODUCTS. There are two alternatives. One is to have the language provide full-screen editing so that BIRTH-

DATA								ENTRY:	

LOCAL VIEW NAME			TYPE	B/OL	FREQ		PERIOD		
department operations									

F	T	DATA ELEMENT NAME	A	FREQ	OCCR	T	LENG	F	FACT
.	.	division	.	....	....	.	....	.	....
.	.	department	.	....	....	.	....	.	....
.	.	manager	.	....	....	.	....	.	....
.	.	employee	.	....	....	.	....	.	....
.	.	skills	.	....	....	.	....	.	....
.	.	salary	.	....	....	.	....	.	....
.	.	work-location	.	....	....	.	....	.	....
.	.	projects	.	....	....	.	....	.	....
.	.	hours-worked	.	....	....	.	....	.	....
.	.	products	.	....	....	.	....	.	....
.	.	description	.	....	....	.	....	.	....
.	.	price	.	....	....	.	....	.	....
.	.	location	.	....	....	.	....	.	....
.	.	. . . .	.	....	....	.	....	.	....
.	.	. . . .	.	....	....	.	....	.	....
.	.	. . . .	.	....	....	.	....	.	....
.	.	. . . .	.	....	....	.	....	.	....
.	.	. . . .	.	....	....	.	....	.	....
.	.	. . . .	.	....	....	.	....	.	....
.	.	. . . .	.	....	....	.	....	.	....
.	.	. . . .	.	....	....	.	....	.	....
.	.	. . . .	.	....	....	.	....	.	....

Figure 22-4. Initial entry of a local view.

DATE can be physically inserted into a meaningful position under EMPLOYEE. The other alternative is to append BIRTHDATE to the bottom of the list and use the "F" and "T" field to define its dependency on EMPLOYEE. Experience indicates that there will be many additional elements added to local views in this manner. One of the things to be studied is the human factors aspects of the two alternatives to determine the preferred approach. The screen in Figure 22-5 illustrates using the "F" and "T" fields in the second alternative.

When the operator has completed specifying the local view, he can hit the ENTER key for preediting. Initial editing will supply association types where unspecified. In this example of minimum keying, none are specified. If an indented element has subordinate elements, shown by further indentation or

DATA								ENTRY:	

LOCAL VIEW NAME			TYPE	B/OL	FREQ		PERIOD

department operations

F	T	DATA ELEMENT NAME	A	FREQ	OCCR	T	LENG	F	FACT
.	.	division	.	....	....	.	....	.	....
.	.	department	.	....	....	.	....	.	....
.	.	manager	.	....	....	.	....	.	....
1	.	employee	.	....	....	.	....	.	....
.	.	skills	.	....	....	.	....	.	....
.	.	salary	.	....	....	.	....	.	....
.	.	work-location	.	....	....	.	....	.	....
.	.	projects	.	....	....	.	....	.	....
.	.	hours-worked	.	....	....	.	....	.	....
.	.	products	.	....	....	.	....	.	....
.	.	description	.	....	....	.	....	.	....
.	.	price	.	....	....	.	....	.	....
.	.	location	.	....	....	.	....	.	....
.	1	birthdate	.	....	....	.	....	.	....
.	.	. . . .	.	....	....	.	....	.	....
.	.	. . . .	.	....	....	.	....	.	....
.	.	. . . .	.	....	....	.	....	.	....
.	.	. . . .	.	....	....	.	....	.	....
.	.	. . . .	.	....	....	.	....	.	....
.	.	. . . .	.	....	....	.	....	.	....
.	.	. . . .	.	....	....	.	....	.	....

Figure 22-5. Adding an additional element.

by the "F" and "T" fields, it is assumed to have a Type M association from its predecessor. Otherwise, it is assumed to have a Type 1 association from its predecessor. It is expected that perhaps 90% of the time these assumptions will be consistent with what the designer has specified on the bubble chart of the local view. The operator must review the supplied association types and make any needed changes. Of course, the operator can explicitly enter the association types when keying in the data names. Hitting the ENTER key results in the screen in Figure 22-6.

In examining the supplied association types, we see that a Type 1 association has been supplied from EMPLOYEE to SKILLS. This is according to the rules since nothing is shown as subordinate to skills. But suppose the designer wants a Type M association here indicating that an employee can have several

DATA   ENTRY:

LOCAL VIEW NAME   TYPE   B/OL   FREQ   PERIOD

DEPARTMENT OPERATIONS

F	T	DATA ELEMENT NAME	A	FREQ	OCCR	T	LENG	F	FACT
		DIVISION	-						
		DEPARTMENT	M						
		MANAGER	1						
1		EMPLOYEE	M						
		SKILLS	1						
		SALARY	1						
		WORK-LOCATION	1						
		PROJECTS	M						
		HOURS-WORKED	1						
		PRODUCTS	M						
		DESCRIPTION	1						
		PRICE	1						
		LOCATION	1						
1		BIRTHDATE	1						

Figure 22-6. Display of system response.

skills listed in the data base. The operator makes this change as shown in Figure 22-7, hits the ENTER key again, and the local view is submitted for full editing. A new screen appears for entry of the next local view. One of the function keys can be used to exit from data entry mode and return to the menu.

DISPLAY EDIT RESULTS

A menu of the type illustrated in Figure 22-8 can be provided for selecting the edit reports of interest. Scrolling and paging should be provided with each display, along with the ability to obtain hard copy output. A single number or a list of numbers, separated by commas, may be specified following the $\Rightarrow$ or an "S" can be keyed before the desired report numbers. Or, a predetermined set of reports can be obtained by keying a one-character code. A function key can cause exit from the display mode back to the menu. Hard copy printing should be available.

EDITING CHANGES

As a result of the input editing, certain changes may be made to the data requirements. Standardizing naming conventions is the major type of change to be made. This includes resolving synomyms and homonyms and also resolving inconsistent associations. Another frequently performed task at this point is assigning standard abreviations to the data names. And then there is always the case in which the designers have changed their minds about something or simply wish to add additional data. The more frequently used editing operations are provided by a series of formatted screens. Any desired editing outside the scope of these screens can be performed by physically changing or adding to the local views. After filling in a screen, the ENTER key causes the screen to be processed and a new screen in the same mode to be presented. A function key can cause processing and an exit from that mode back to the menu. The menu in Figure 22-9 takes the operator into the input editing mode.

Equate Data Names

The EQUATE screen (Figure 22-10) is used to assign a standard name to be used in place of the several variants that have been entered on various local views. Using this function once avoids the necessity of physically changing the several views. This function can also be used to specify abbreviations for data names.

This screen filled in by the operator would appear as in Figure 22-11.

DATA   ENTRY:

LOCAL VIEW NAME   TYPE   B/OL   FREQ   PERIOD

DEPARTMENT OPERATIONS

F	T	DATA ELEMENT NAME	A	FREQ	OCCR	T	LENG	F	FACT
		DIVISION	-						
		DEPARTMENT	M						
		MANAGER	1						
	1	EMPLOYEE	M						
		SKILLS	m						
		SALARY	1						
		WORK-LOCATION	1						
		PROJECTS	M						
		HOURS-WORKED	1						
		PRODUCTS	M						
		DESCRIPTION	1						
		PRICE	1						
		LOCATION	1						
	1	BIRTHDATE	1						

Figure 22-7. Changing the EMPLOYEE-SKILLS association type.

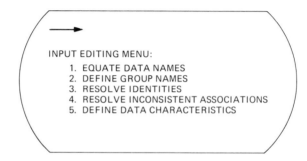

```
DISPLAY EDIT REPORTS:
    1. EDITED LOCAL VIEWS
    2. DATA DEFINITION AND WHERE-USED LIST
    3. KEYWORD-IN-CONTEXT (KWIC) LIST
    4. INCONSISTENT ASSOCIATIONS
    5. INTERSECTING ATTRIBUTES
    6. NEW ELEMENT REPORT
    7. SOURCE-USE EXCEPTIONS
    8. ALL
```

Figure 22-8. Menu for displaying edit reports.

```
INPUT EDITING MENU:
    1. EQUATE DATA NAMES
    2. DEFINE GROUP NAMES
    3. RESOLVE IDENTITIES
    4. RESOLVE INCONSISTENT ASSOCIATIONS
    5. DEFINE DATA CHARACTERISTICS
```

Figure 22-9. Input editing menu.

```
EQUATE DATA NAMES:
NAME:                           ABBREVIATION:
SYNONYM(S):
```

Figure 22-10. Screen for equating data element names.

Define Group Names

The GROUP function is used to specify that several related names are to be structured into one field and to define a group name for that field. The field can be referred to by its new group name or by any of its component names. The screen in Figure 22-12 is presented to the operator.

When filled in by the operator, the screen may appear as in Figure 22-13.

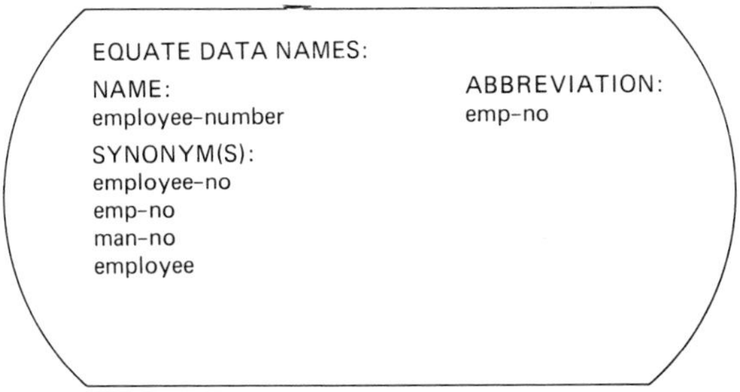

EQUATE DATA NAMES:

NAME: ABBREVIATION:
employee-number   emp-no

SYNONYM(S):
employee-no
emp-no
man-no
employee

Figure 22-11. Example of equating data element names.

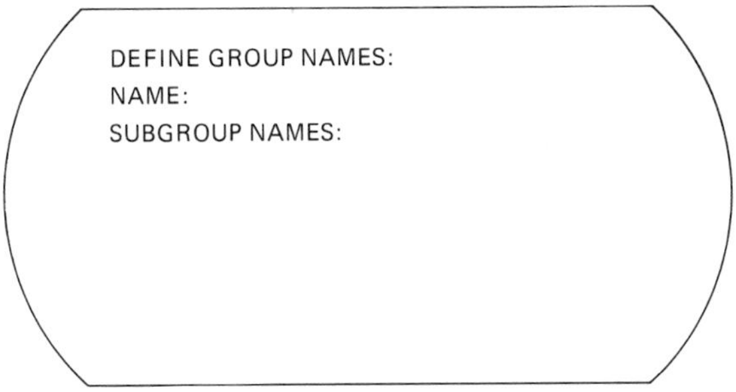

DEFINE GROUP NAMES:
NAME:
SUBGROUP NAMES:

Figure 22-12. Screen for defining group names.

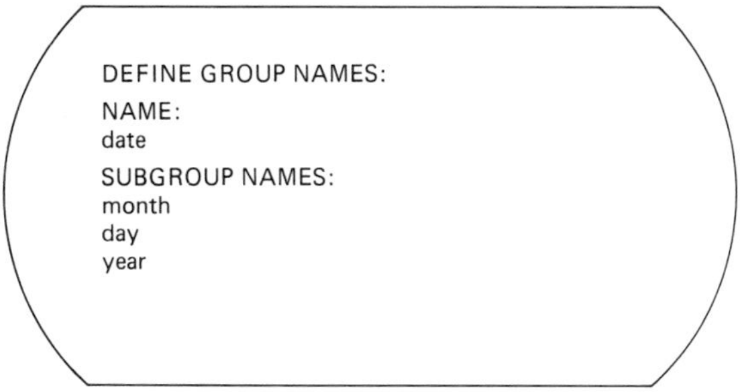

DEFINE GROUP NAMES:
NAME:
date
SUBGROUP NAMES:
month
day
year

Figure 22-13. Example of defining group names.

Resolve Identities

An identity is when two data elements each point to the other with a Type 1 association. The screen in Figure 22-14 is presented to the operator.

There are several ways of resolving identities. One way is to implement both elements in the same segment with one as the primary key and the other as an

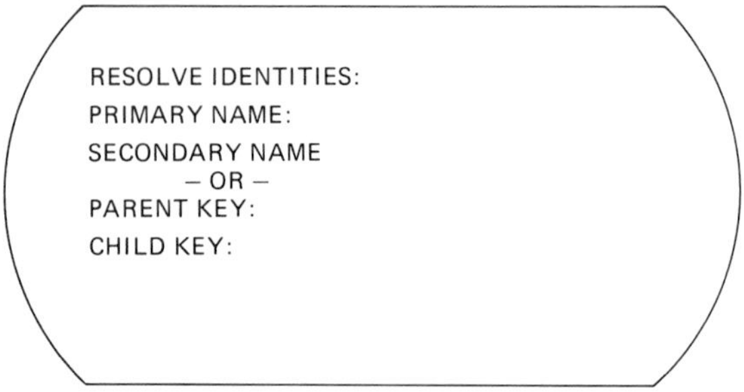

```
RESOLVE IDENTITIES:
PRIMARY NAME:
SECONDARY NAME
        — OR —
PARENT KEY:
CHILD KEY:
```

Figure 22-14. Screen for resolving identities.

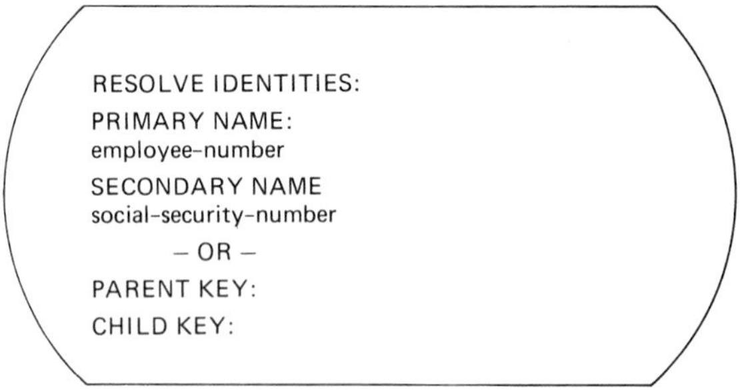

```
RESOLVE IDENTITIES:
PRIMARY NAME:
employee-number
SECONDARY NAME
social-security-number
        — OR —
PARENT KEY:
CHILD KEY:
```

Figure 22-15. Example of resolving identities (1).

attribute that will be a candidate for secondary indexing. For this solution, the completed screen may appear as in Figure 22-15.

Alternately, perhaps for security reasons, the designer may wish to resolve the identity by implementing the two elements as keys of two segments, one as the physical child of the other. This solution is illustrated in Figure 22-16.

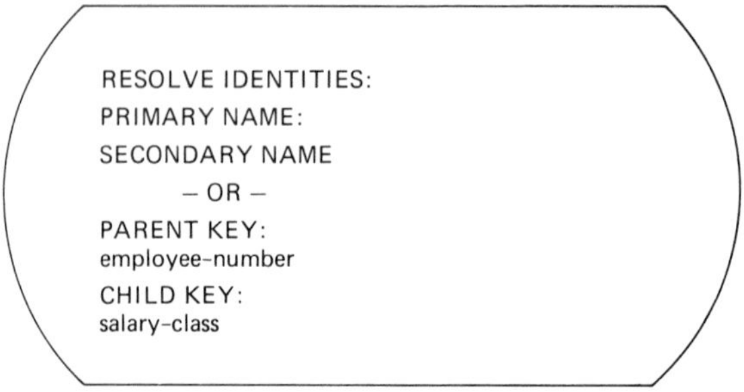

```
RESOLVE IDENTITIES:
PRIMARY NAME:
SECONDARY NAME
        — OR —
PARENT KEY:
employee-number
CHILD KEY:
salary-class
```

Figure 22-16. Example of resolving identities (2).

Inconsistent Associations

If the designer does not resolve inconsistent associations by creating two distinct target elements, he must resolve them by specifying the association type to be used in the design. The screen in Figure 22-17 is presented to the operator.

```
RESOLVE INCONSISTENT ASSOCIATIONS:
"FROM" ELEMENT:
"TO" ELEMENT:
ASSOCIATION TYPE:
"TO" ELEMENT:
ASSOCIATION TYPE:
```

Figure 22-17. Screen for resolving inconsistent associations.

Suppose that the association from CUSTOMER to ADDRESS has been specified in one case as Type 1 and in another case as Type M. Upon investigating, the designer learns that in one case the function is working with shipping address (there may be several), and in another case the function is working with billing address (there is one). In the screen in Figure 22-18, he

```
RESOLVE INCONSISTENT ASSOCIATIONS:
"FROM" ELEMENT:
customer-number
"TO" ELEMENT:
customer-address
ASSOCIATION TYPE:
M
"TO" ELEMENT:
ASSOCIATION TYPE:
```

Figure 22-18. Example of resolving inconsistent associations (1).

specifies that there will be only one set of addresses and the Type M association will be used. This decision can have a significant impact on the application program logic.

If, on the other hand, the designer decides to implement two sets of addresses, the screen may be filled out as in Figure 22-19.

RESOLVE INCONSISTENT ASSOCIATIONS:

"FROM" ELEMENT:
customer-number

"TO" ELEMENT:
shipping-address

ASSOCIATION TYPE:
M

"TO" ELEMENT:
billing-address

ASSOCIATION TYPE:
1

Figure 22-19. Example of resolving inconsistent associations (2).

Define Data Characteristics

Frequently, the optional data characteristics are not known at the beginning of a design study and it is desirable to enter them at some later stage. Also, once entered and edited, inconsistencies should be resolved. Data characteristics may be entered with initial data entry, but if not entered at that time, they can be entered separately. For separate entry, the screen in Figure 22-20 is used.

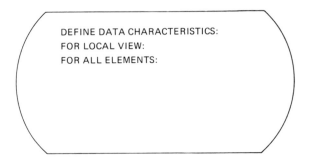

DEFINE DATA CHARACTERISTICS:
FOR LOCAL VIEW:
FOR ALL ELEMENTS:

Figure 22-20. Screen for defining data characteristics.

The operator specifies whether he wants to enter data characteristics for elements of a particular local view or for the entire collection of data elements. In the screen in Figure 22-21, a local view is selected.

After making this specification, the screen in Figure 22-22 is presented. All data element names of the selected category are listed in alphabetical order.

The operator then enters data characteristics for any or all of the elements shown as in Figure 22-23.

DEFINE DATA CHARACTERISTICS:

FOR LOCAL VIEW:
department operations

FOR ALL ELEMENTS:

Figure 22-21. Example of specifying data characteristics for a local view.

DEFINE DATA CHARACTERISTICS FOR DEPARTMENT OPERATIONS

DATA ELEMENT NAME	FREQ	OCCR	T	LENG	F	FACT
BIRTHDATE	..	..	.	..	.	..
EMPLOYEE	..	..	.	..	.	..
SALARY	..	..	.	..	.	..
SKILLS	..	..	.	..	.	..
WORK-LOCATION	..	..	.	..	.	..
....	..	..	.	..	.	..
....	..	..	.	..	.	..
....	..	..	.	..	.	..
....	..	..	.	..	.	..
....	..	..	.	..	.	..
....	..	..	.	..	.	..
....	..	..	.	..	.	..
....	..	..	.	..	.	..
....	..	..	.	..	.	..
....	..	..	.	..	.	..
....	..	..	.	..	.	..

Figure 22-22. Screen for entering data characteristics.

DISPLAY STRUCTURING DIAGNOSTICS

Although the structuring diagnostics are of utmost importance to the designer, there are no unique features in the suggested language for displaying these results. A simple menu for selecting the desired reports can be provided. The menu can be of the type illustrated in Figure 22-24 for selecting design reports. Scrolling and paging should be provided with each display, along with the ability to obtain hard copy output.

DEFINE DATA CHARACTERISTICS FOR DEPARTMENT OPERATIONS

DATA ELEMENT NAME	FREQ	OCCR	T	LENG	F	FACT
BIRTHDATE	..	.. 1	n	. 5	.	..
EMPLOYEE	..	. 50	a	.15	.	..
SALARY	..	.. 1	n	. 5	.	..
SKILLS	..	. 10	a	. 8	.	..
WORK-LOCATION	..	.. 5	a	. 5	.	..
....	..	....	.	...	.	..
....	..	....	.	...	.	..
....	..	....	.	...	.	..
....	..	....	.	...	.	..
....	..	....	.	...	.	..
....	..	....	.	...	.	..
....	..	....	.	...	.	..
....	..	....	.	...	.	..
....	..	....	.	...	.	..
....	..	....	.	...	.	..
....	..	....	.	...	.	..
....	..	....	.	...	.	..

Figure 22-23. Example of entering data characteristics.

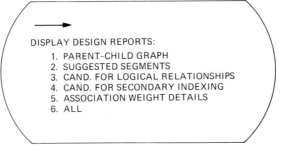

DISPLAY DESIGN REPORTS:
  1. PARENT-CHILD GRAPH
  2. SUGGESTED SEGMENTS
  3. CAND. FOR LOGICAL RELATIONSHIPS
  4. CAND. FOR SECONDARY INDEXING
  5. ASSOCIATION WEIGHT DETAILS
  6. ALL

Figure 22-24. Menu for displaying design reports.

DISPLAY DESIGN RESULTS

The menu in Figure 22-24 is used to select the design reports to be viewed. A single number or a list of numbers, separated by commas, may be specified following the → or by keying an "S" before desired report numbers. A predetermined set of reports can be requested by keying a one-character code. Scrolling and paging can be provided with each selected report. A function key can cause exit from the display mode back to the menu. Hard copy printing will be available.

STRUCTURE REFINEMENT

There are a number of language and display features that can assist the designer in evaluating and refining the logical design. One desirable feature would be the display of the logical data base, derived from the designed physical data bases, that will support a given application or function. Another desirable feature would be the ability to display the names, or the bubble charts, of the local views that participated in the structuring of a given portion of the logical model. Using structure specification commands suggested in Chapter 12, the designer can interactively specify refinements and then observe the results (including diagnostics).

Using a split-screen capability, the dual display concept can be extended to real-time interactive refinement of the design. Referring to the screen layout in Figure 22-25, selected local views or even the composite model can be dis-

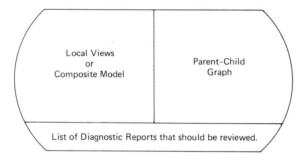

Figure 22-25. Concept of split-screen display of data requirements and resulting logical design.

played with the resulting logical model. A "keys-only" subset of the composite model will save display space and in many cases will be sufficient to define the logical model. By a light pen or keystrokes, the designer can change, delete, or create portions of the local view of the composite model on the left and witness the resulting structural implications on the right. A zoom lens can enlarge or reduce the size of the image on either side of the screen to help the designer focus on the "forest" or on the "trees." At the bottom of the screen a list of diagnostic reports containing interesting information can be displayed for each refinement. If more display space is needed than is available on a single screen, two or more display terminals could be used.

APPENDICES
APPENDIX A: CONCEPTUAL AND LOGICAL DESIGN CASE STUDY
APPENDIX B: PROFESSIONAL READER'S REVIEW AND ANALYSIS

Appendix A: Conceptual and Logical Design Case Study

INTRODUCTION

A case study embodying many of the concepts of computer-assisted data base design will be presented. Emphasis is given to the iterative nature of the process and to the control of the process by the human designer acting in response to the information presented in the edit, diagnostic, and design reports. The iterations will be carried through the conceptual and logical design phases. The application of physical design procedures will be suggested, but will not be illustrated in detail.

OBJECTIVE

This case study presents an illustrative, but not exhaustive, example of using the automated techniques that have been presented. The example is necessarily small in size, and many factors necessary to a real design study, such as update requirements and data element sizes and occurrences for space calculations, are omitted. The overall objective is to illustrate the major features of computer-assisted data base design as applied to conceptual and logical design. In so doing, an attempt has been made to demonstrate:

- The logical design structures derived by the automated procedures
- A sampling of the editing and diagnostic information that can be provided
- An indication of intelligent constraints automatically applied to the structuring process
- The role of the human designer in reacting to this information and in controlling the overall process

In the analyses of each iteration, an attempt will be made to present a representative example of things the designer should consider and to indicate when reports from the automated procedures can provide information helpful to these considerations.

DESCRIPTION

This case study involves the design of a DL/I data base for a football league. The data base will be central to the league itself and will not be for the individual teams. The league directors want to produce reports that will provide records of the games that have been played and of the players' performances. By measuring the players' performances, the directors hope both to gain a measure of the balance among the several teams and to decide if the players are being fairly paid. Information about the number of penalties and the number of injuries sustained by the players is also wanted in the reports. With this information the directors can take action, as necessary, to keep the games as clean and safe as possible.

The league has 30 teams, and during each week of the season 15 games are played. Each team has 60 players and 6 coaches. On the average, 40 players on a team participate in a game, each player plays 1.2 positions, and 5 players receive injuries of some sort.

This case study will be presented in six iterations:

Iteration 1: Conceptual design —Editing
Iteration 2: Conceptual design —Reediting
Iteration 3: Logical design   —Initial structuring
Iteration 4: Logical design   —Refinement
Iteration 5: Logical design   —Adding additional requirements
Iteration 6: Logical design   —Adding future requirements

Some final comments will then be made regarding alternative logical designs that can be further evaluated in the physical design phase.

For the first two iterations, the design process will not be carried beyond the editing procedures of the conceptual design phase. The diagnostic and design reports of the logical design phase are usually not fully meaningful until the inconsistencies in data names and association types have been resolved. Indeed, one of the biggest mistakes when using computer-assisted techniques is the tendency of looking at the design reports too early in the process and becoming confused by unusual results caused by problems that should have been resolved before passing from the conceptual design phase to the logical design phase.

ITERATION 1 CONCEPTUAL DESIGN—EDITING

Initial Data Requirements

The initial data base design is to support the player performance and the game record applications which together consist of nine functions. These are considered to be the most important of the applications that will be implemented. Six of these functions are batch functions and the other three functions are online queries. For purposes of simplicity, the data requirements of these functions will be derived from their expected outputs, although in a real design study inputs and special processing requirements, if any, should also be used.

The external view of each function is presented by portraying a brief verbal description of the function with the format and content of the expected output. The binary relationships of its local view (i.e., bubble charts), derived by the designer or end user or both from the output descriptions, are also included. For our starting point, we will assume that these local views have already been derived through dialogues between the designer and end user.

The reader will note that the designer and end user have not done a very good job of choosing and standardizing the data element names. In a real design study we would like for them to do better, but, had they done so for this case study, the capabilities of the editing procedures could not be as well demonstrated.

Descriptions of the nine external views follow in Figures A-1 through A-9.

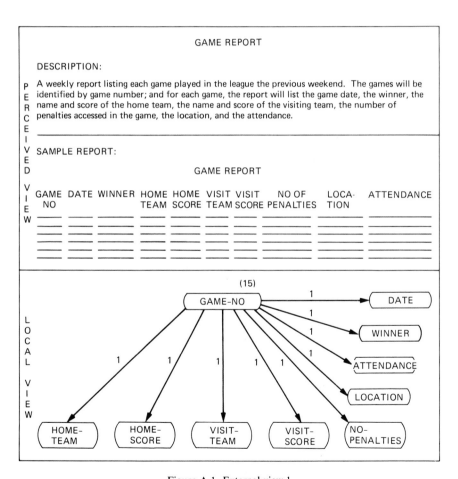

Figure A-1. External view 1.

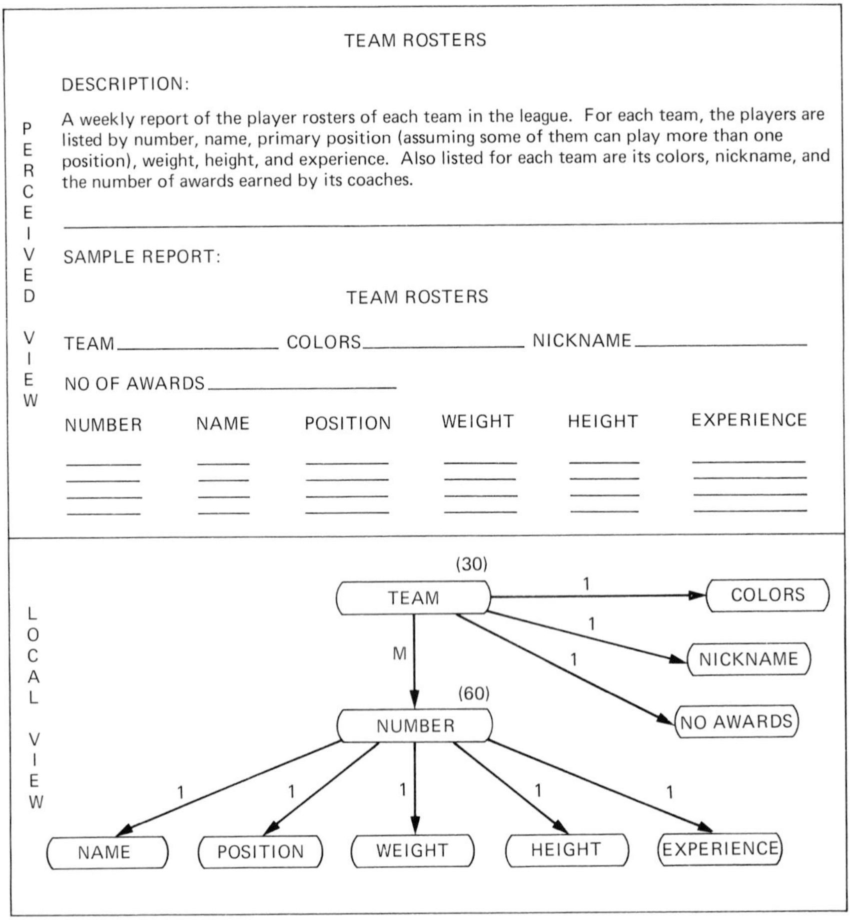

Figure A-2. External view 2.

Reported Results

The editing procedures combine the binary relations of the nine local views into a composite model, and the major editing reports proposed in Chapter 5 are produced. The edited local view reports are not illustrated in this case study because they are concerned mostly with the mechanical errors of entering binary relations rather than with errors associated with interpreting the applications' data requirements. The editing reports of primary interest are:

- Keyword-in-Context (KWIC) list
- Inconsistent Associations report
- Intersecting Attributes report

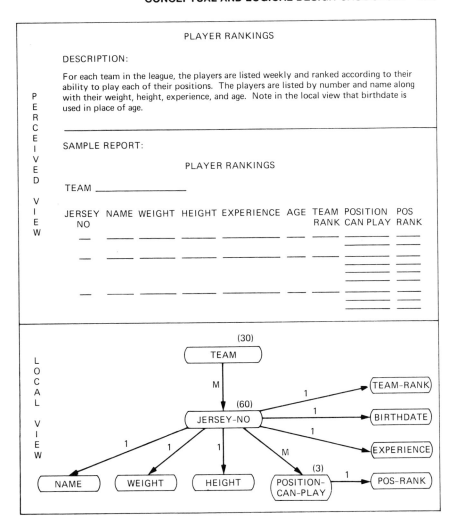

PERCEIVED VIEW

PLAYER RANKINGS

DESCRIPTION:

For each team in the league, the players are listed weekly and ranked according to their ability to play each of their positions. The players are listed by number and name along with their weight, height, experience, and age. Note in the local view that birthdate is used in place of age.

SAMPLE REPORT:

PLAYER RANKINGS

TEAM _____

| JERSEY NO | NAME | WEIGHT | HEIGHT | EXPERIENCE | AGE | TEAM RANK | POSITION CAN PLAY | POS RANK |

LOCAL VIEW

Figure A-3. External view 3.

While reports of these types can reveal a surprising number of editing problems, they cannot disclose all of the problems that may exist. The following examples illustrate the kinds of editing problems that can be revealed.

Keyword-in-Context (KWIC) List

A number of naming problems are evident from the KWIC list in Figure A-10. We see an unqualified DATE and a BIRTHDATE. Does DATE refer to birthdate also or to some other kind of date? NAME and PLAYER-NAME

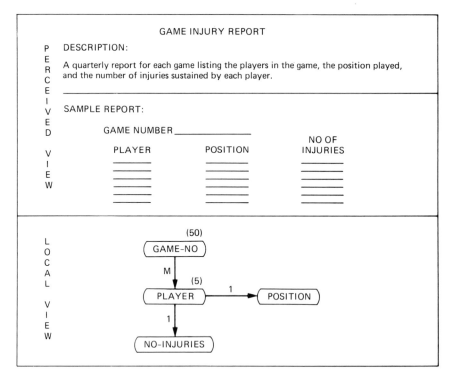

Figure A-4. External view 4.

give rise to the same kind of question. In observing JERSEY-NO and PLAYER-NO, one would suspect that these are two names for the same data. Moving down, does PLAYER refer to the number or name of the player? And for POSITION, a problem of interpretation is suggested by its three uses. Does the end user mean the one position at which the player is currently assigned or the several positions at which he can play or the several positions at which he has played?

Other problems, not quite so evident, can also be observed. The designer notices the names SCORE, HOME-SCORE, and VISIT-SCORE. Are three separate data elements required or will one or two suffice? The same question can be asked of TEAM, HOME-TEAM, VISIT-TEAM, OPPONENT, and WINNER.

Inconsistent Associations Report

No inconsistent associations were detected in this iteration.

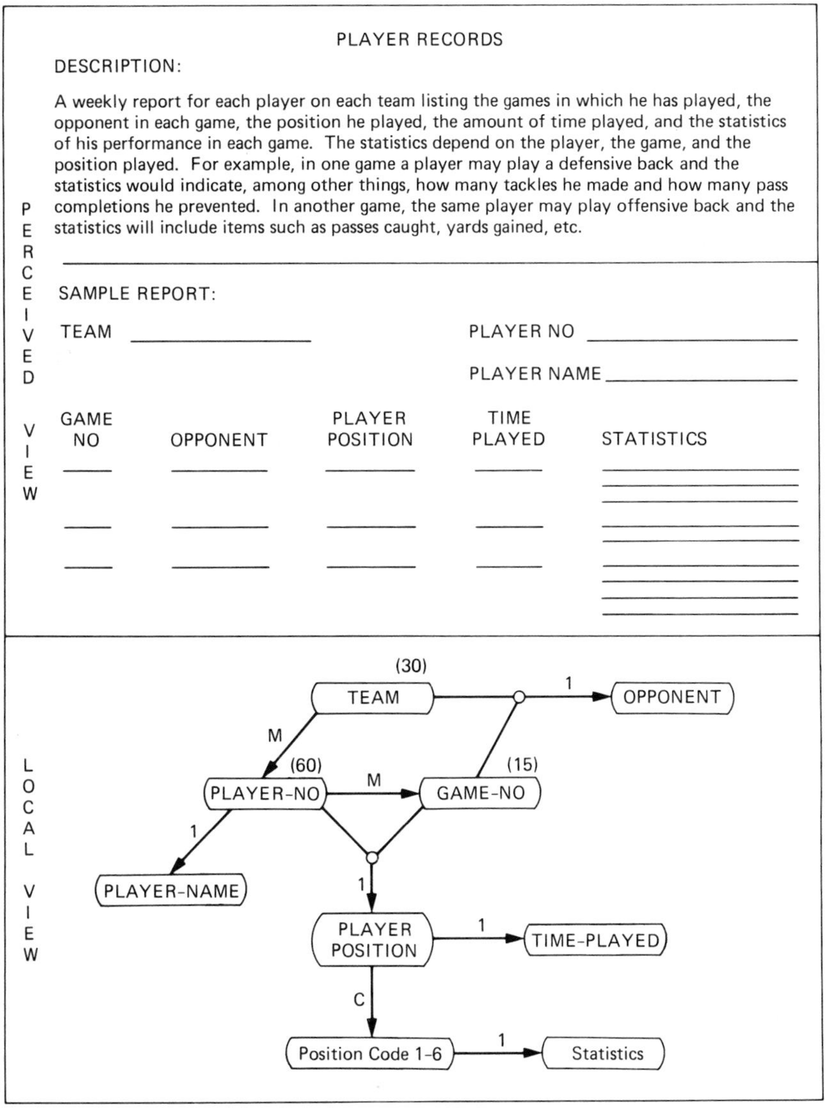
PLAYER RECORDS
DESCRIPTION:
A weekly report for each player on each team listing the games in which he has played, the opponent in each game, the position he played, the amount of time played, and the statistics of his performance in each game. The statistics depend on the player, the game, and the position played. For example, in one game a player may play a defensive back and the statistics would indicate, among other things, how many tackles he made and how many pass completions he prevented. In another game, the same player may play offensive back and the statistics will include items such as passes caught, yards gained, etc.
PERCEIVED VIEW
SAMPLE REPORT:
TEAM
PLAYER NO
PLAYER NAME
GAME NO
OPPONENT
PLAYER POSITION
TIME PLAYED
STATISTICS
LOCAL VIEW
(30)
TEAM
1
OPPONENT
M
(60)
PLAYER-NO
M
(15)
GAME-NO
1
PLAYER-NAME
1
PLAYER POSITION
1
TIME-PLAYED
C
Position Code 1-6
1
Statistics

Figure A-5. External view 5.

Intersecting Attributes Report

The Intersecting Attributes report in Figure A-11 displays some of the naming problems already observed in the KWIC list with some additional revelations. NUMBER appears also to be a synonym of JERSEY-NO because both have

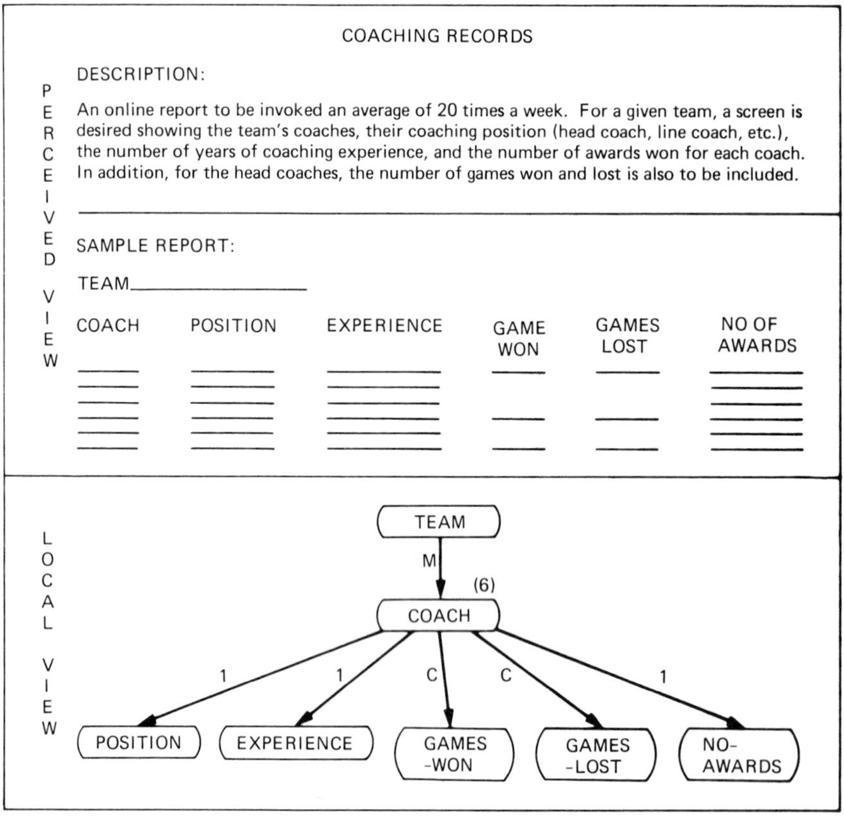
COACHING RECORDS
PERCEIVED VIEW
DESCRIPTION:
An online report to be invoked an average of 20 times a week. For a given team, a screen is desired showing the team's coaches, their coaching position (head coach, line coach, etc.), the number of years of coaching experience, and the number of awards won for each coach. In addition, for the head coaches, the number of games won and lost is also to be included.
SAMPLE REPORT:
TEAM
COACH
POSITION
EXPERIENCE
GAME WON
GAMES LOST
NO OF AWARDS
LOCAL VIEW
TEAM
M
(6)
COACH
1
1
C
C
1
POSITION
EXPERIENCE
GAMES -WON
GAMES -LOST
NO- AWARDS

Figure A-6. External view 6.

been used as keys of EXPERIENCE. COACH is also a key of EXPERIENCE. Are there two kinds of experience?

Attention is also focused on NO-AWARDS for COACH and for TEAM. Does NO-AWARDS pertain to the number of awards earned by each coach on the team? If so, is NO-AWARDS for the team intended to be the total number of awards earned by the team's coaches, or does it refer to special team awards that are different from coaches' awards? The same types of questions can be asked about NO-PENALTIES with regard to players and the team as a whole.

Analysis of Results

As a result of the naming inconsistencies suggested by the editing reports of this iteration, the data base designer can go to the end user with specific ques-

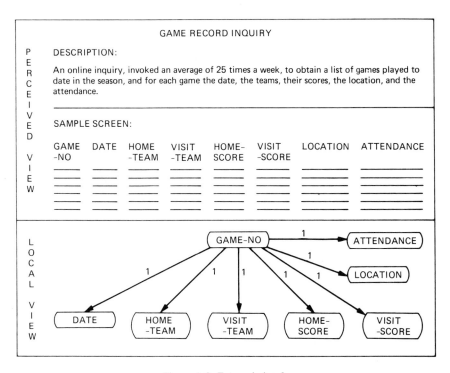

TEAM RECORD INQUIRY

PERCEIVED VIEW

DESCRIPTION:

An online inquiry to be invoked an average of 15 times a week. For a given team in a given game the screen shows the name of the opposing team, the winner and the score of the game.

SAMPLE SCREEN:

TEAM_____   GAME-NO_____

OPPONENT_____   WINNER_____

SCORE_____

LOCAL VIEW

Figure A-7. External view 7.

GAME RECORD INQUIRY

PERCEIVED VIEW

DESCRIPTION:

An online inquiry, invoked an average of 25 times a week, to obtain a list of games played to date in the season, and for each game the date, the teams, their scores, the location, and the attendance.

SAMPLE SCREEN:

| GAME -NO | DATE | HOME -TEAM | VISIT -TEAM | HOME- SCORE | VISIT -SCORE | LOCATION | ATTENDANCE |

LOCAL VIEW

Figure A-8. External view 8.

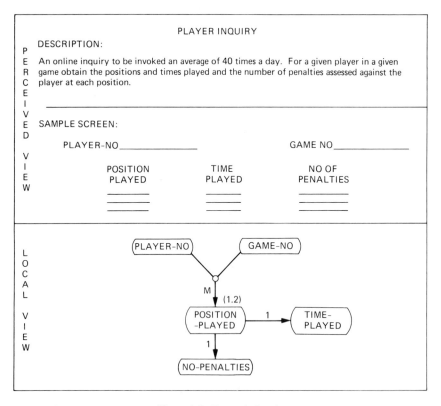

Figure A-9. External view 9.

tions and specific recommendations regarding the interpretation and renaming of many of the elements.

The designer and the end user determine that the awards for a team are different from the coaches' awards, and so they choose to create two different awards fields: NO-TEAM-AWARDS and NO-COACHING-AWARDS. They also determine that the number of penalties for a team is the sum of the number of penalties for the players in a game, and it will be called TOTAL-NO-PENALTIES. There is also a question as to whether this summary field should be included in the data base or if it can be calculated. This latter question will be addressed in the analysis of Iteration 3 after looking at the suggested logical design.

Players are identified consistently by PLAYER-NO, and PLAYER-NAME is used instead of NAME. GAME-DATE replaces DATE. POSITION becomes PRIMARY-POSITION and POSITION-PLAYED, as appropriate, for players, and COACHING-POSITION for coaches. EXPERIENCE is similarly qualified between players and coaches.

```
KEYWORD-IN-CONTEXT (KWIC) LIST
                    *
                    *
                    DATE
        BIRTH-      DATE
                    *
                    *
                    NAME
        PLAYER-     NAME
                    *
                    *
        JERSEY-     NO
        PLAYER-     NO
                    *
                    *
                    OPPONENT
                    *
                    *
                    PLAYER
                    PLAYER      -NAME
                    PLAYER      -NO
                    PLAYER      -POSITION
                    *
                    *
                    POSITION
                    POSITION    -CAN-PLAY
                    POSITION    -PLAYED
        PLAYER-     POSITION
                    *
                    *
                    SCORE
        HOME-       SCORE
        VISIT-      SCORE
                    *
                    *
                    TEAM
        HOME-       TEAM
        VISIT-      TEAM
                    *
                    *
                    WINNER
                    *
                    *
```

Figure A-10. Keyword-in-Context (KWIC) list, iteration 1.

INTERSECTING ATTRIBUTES REPORT	
Key	Attribute
*	
*	
COACH	EXPERIENCE
JERSEY-NO	"
NUMBER	"
JERSEY-NO	HEIGHT
NUMBER	"
JERSEY-NO	NAME
NUMBER	"
COACH	NO-AWARDS
TEAM	"
	NO.-
GAME-NO	PENALTIES
POSITION-PLAYED	"
COACH	POSITION
PLAYER	"
NUMBER	"
PLAYER-POSITION	TIME-PLAYED
POSITION-PLAYED	"
JERSEY-NO	WEIGHT
NUMBER	"
*	
*	

Figure A-11. Intersecting Attributes report, iteration 1.

The designer decides to defer the resolution of certain other editing problems, thinking that the logical model, when it can be more meaningfully obtained, will help him decide among various alternative resolutions. Thus, the resolution of SCORE, HOME-SCORE, and VISIT-SCORE and of TEAM, HOME-TEAM, VISIT-TEAM, OPPONENT, and WINNER will be deferred to a later iteration.

ITERATION 2 CONCEPTUAL DESIGN—REEDITING

Revised Data Requirements

Assuming that the dialogues indicated above have taken place, a number of data element naming changes will be made to seven of the nine local views.

The name changes to be made are indicated in the revised local views shown in Figures A-12 through A-18. External views 7 and 9 remain unchanged. Because these changes are rather extensive and may create additional problems, another editing iteration will be made before proceeding to logical design.

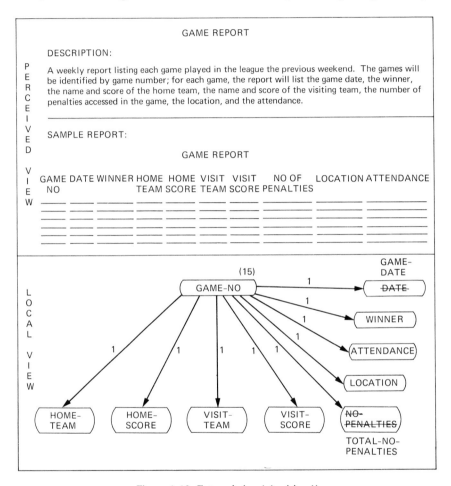

Figure A-12. External view 1 (revision 1).

Reported Results

From this second iteration, the KWIC Inconsistent Associations, and Intersecting Attributes reports are obtained as before, and some new editing problems are revealed in addition to those that were purposely left unresolved. The relevant portions of these reports are shown below.

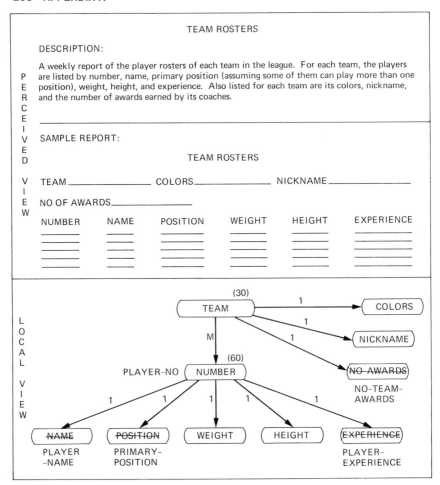

Figure A-13. External view 2 (revision 1).

Keyword-in-Context List

Although there are some new entries on this report, no new problems are revealed. Thus, the report is not reproduced here. Refer to Figure A-10.

Inconsistent Associations Report

An Inconsistent Associations report was produced in this iteration and is shown in Figure A-19. Inconsistent associations are reported from the compound key, **GAME-NO*PLAYER-NO**, to **POSITION-PLAYED**. In external view 5, the

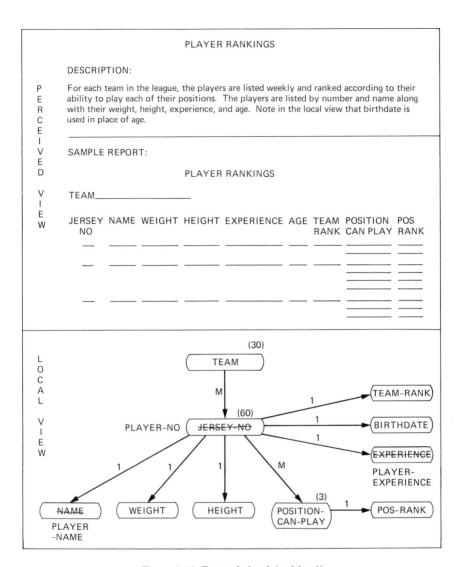

Figure A-14. External view 3 (revision 1).

association was specified as Type 1; in external view 9, it was specified as Type M.

Intersecting Attributes Report

The Intersecting Attributes report in Figure A-20 shows that PLAYER-NO and PLAYER-NAME are both used as keys for PRIMARY-POSITION. In the other local views that pertain to players, PLAYER-NO is used as an identifier of PLAYER-NAME.

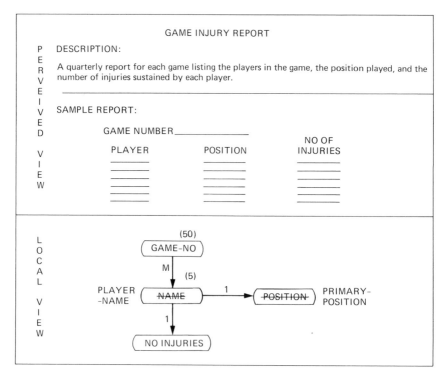

Figure A-15. External view 4 (revision 1).

Analysis of Results

In response to questions from the designer regarding POSITION-PLAYED, the end user specifies that what he really wants in both cases are the several positions a player may have played in a given game. Thus, the designer will change external view 5 to show a Type M association between GAME-NO*PLAYER-NO and POSITION-PLAYED.

The designer and the end user also decide to add PLAYER-NO to external view 4 to be used as the identifier of a given player, and to let PLAYER-NAME be an attribute of PLAYER-NO. In so doing, they are using these names in a manner that is consistent with the other external views.

ITERATION 3 LOGICAL DESIGN—INITIAL STRUCTURING

Revised Data Requirements

The changes determined in iteration 2 are made to external views 4 and 5 and are illustrated in Figures A-21 and A-22. Since these seem to be minor

PLAYER RECORDS

DESCRIPTION:

A weekly report for each player on each team listing the games in which he has played, the opponent in each game, the position he played, the amount of time played, and the statistics of his performance in each game. The statistics depend on the player, the game, and the position played. For example, in one game a player may play a defensive back and the statistics would indicate, among other things, how many tackles he made and how many pass completions he prevented. In another game, the same player may play offensive back and the statistics will include items such as passes caught, yards gained, etc.

P
E
R
C
E   SAMPLE REPORT:
I
V   TEAM _____   PLAYER NO_____
E
D   PLAYER NAME_____

V   GAME   PLAYER –   TIME
I   NO   OPPONENT   POSITION   PLAYED   STATISTICS
E   ___   _____   _____   _____   _____
W   _____

     ___   _____   _____   _____   _____
                                                _____
     ___   _____   _____   _____   _____
                                                _____
                                                _____

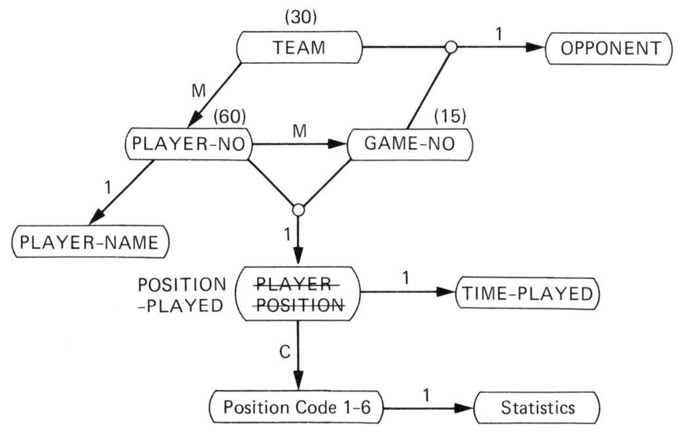

Figure A-16. External view 5 (revision 1).

changes, the designer chooses to carry this third iteration through the logical design phase.

Reported Results

Having carried this iteration through the logical design phase, we are interested in the design and diagnostic reports as well as the edit reports. To con-

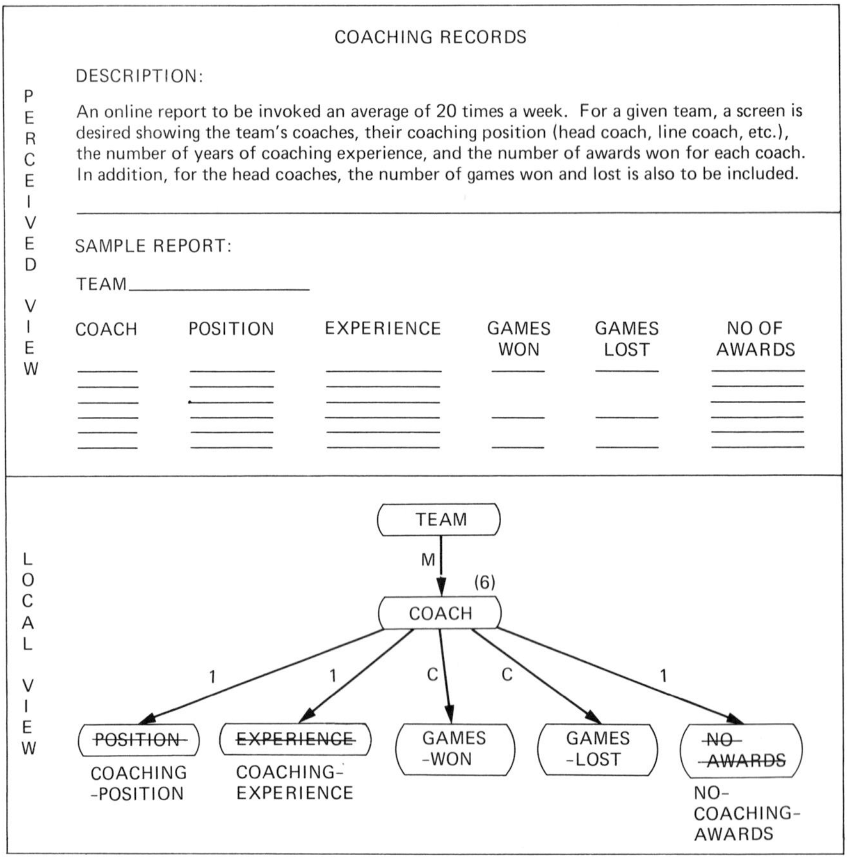

Figure A-17. External view 6 (revision 1).

serve space, only the portions of these reports that are relevant to analyzing the results of this iteration will be displayed. As before, a brief analysis of the results will follow the reports.

Edit Reports

As expected, the editing problems are all resolved except those purposely deferred. Thus, the editing reports for this iteration are not illustrated.

Design Reports

The design reports from these procedures are:

- Parent-Child graph
- Suggested Segments report

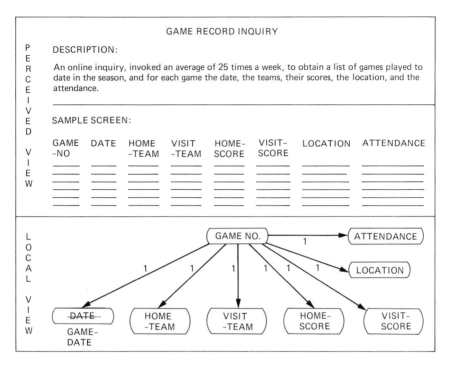

Figure A-18. External view 8 (revision 1).

INCONSISTENT ASSOCIATIONS REPORT

From Element	To Element	Assoc Type	Ext View
GAME-NO*PLAYER-NO	POSITION-PLAYED	1	EV-5
GAME-NO*PLAYER-NO	POSITION-PLAYED	M	EV-9

Figure A-19. Inconsistent Associations report, iteration 2.

- Candidates for Logical Relationships
- Candidates for Secondary Indexing

Each of these reports resulting from this iteration are represented and discussed below.

Parent-Child Graph

The segments and their physical relationships are depicted in Figure A-23. The names inside the segment boxes are the names of the key fields because seg-

INTERSECTING ATTRIBUTES REPORT

Key	Attribute
*	
*	
PLAYER-NAME	PRIMARY-POSITION
PLAYER-NO.	,,
*	
*	

Figure A-20. Intersecting Attributes report, iteration 2.

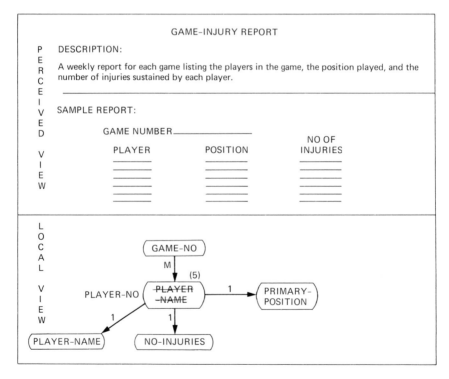

Figure A-21. External view 4 (revision 2).

PERCEIVED VIEW

PLAYER RECORDS

DESCRIPTION:

A weekly report for each player on each team listing the games in which he has played, the opponent in each game, the position he played, the amount of time played, and the statistics of his performance in each game. The statistics depend on the player, the game, and the position played. For example, in one game a player may play a defensive back and the statistics would indicate, among other things, how many tackles he made and how many pass completions he prevented. In another game, the same player may play offensive back and the statistics will include items such as passes caught, yards gained, etc.

SAMPLE REPORT:

TEAM ____________ PLAYER NO ____________

PLAYER NAME ____________

GAME NO	OPPONENT	PLAYER POSITION	TIME PLAYED	STATISTICS
____	____	____	____	____
____	____	____	____	____
____	____	____	____	____

LOCAL VIEW

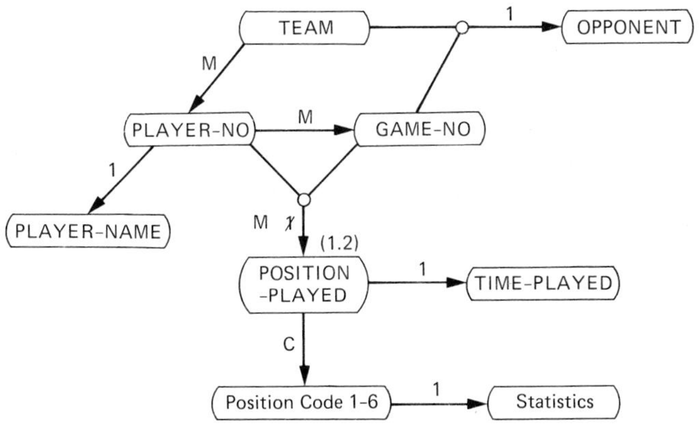

Figure A-22. External view 5 (revision 2).

PARENT-CHILD GRAPH

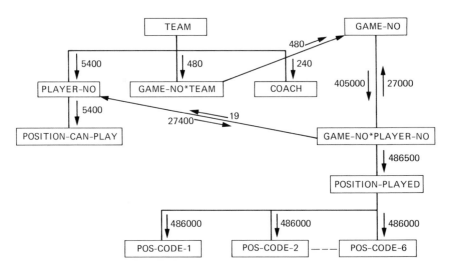

Figure A-23. Parent-Child graph, iteration 3.

ment names have not yet been assigned. Candidates for logical relations are indicated by diagonal lines; for the intersection segments, the choice of physical parent was made according to the performance weights. The children of each parent are in left-to-right order according to performance weights. The performance weights, calculated for a setup period of one week as described in Chapter 10, are shown as numeric values with an arrow indicating the direction of traversal.

Suggested Segments Report

The Suggested Segments report, illustrated in Figure A-24, shows the key field of each segment and all the attributes (i.e., nonkey fields) assigned to each key's segment. Also shown is the performance weight from the key to each of its attributes.

Candidates for Logical Relations Report

The Candidates for Logical Relations report in Figure A-25 shows all the places in the logical model where the automated procedures have determined that logical relations should be considered. The elements to be considered as parents and intersections are listed.

SUGGESTED SEGMENTS		
Key	Attribute	Weight
COACH	COACHING-EXPERIENCE	240
	COACHING-POSITION	240
	GAMES-LOST	240
	GAMES-WON	240
	NO-COACHING-AWARDS	240
GAME-NO	ATTENDANCE	65
	GAME-DATE	65
	HOME-SCORE	65
	HOME-TEAM	65
	LOCATION	65
	VISIT-SCORE	65
	VISIT-TEAM	65
	WINNER	45
	SCORE	30
	TOTAL-NO-PENALTIES	15
PLAYER-NO	PLAYER-NAME	5419
	HEIGHT	3600
	PLAYER-EXPERIENCE	3600
	WEIGHT	3600
	PRIMARY-POSITION	1819
	BIRTHDATE	1800
	TEAM-RANK	1800
	NO-INJURIES	19
POS-CODE-1 through POS-CODE-6	Attributes not listed to conserve space	486000
POSITION-CAN-PLAY	POS-RANK	5400
POSITION-PLAYED	TIME-PLAYED	486500
	NO-PENALTIES	480
TEAM	COLORS	30
	NICKNAME	30
	NO-TEAM-AWARDS	30
GAME-NO*PLAYER-NO	None	
GAME-NO*TEAM	OPPONENT	480

Figure A-24. Suggested segments, iteration 3.

Candidates for Secondary Indexing

The Candidates for Secondary Indexing report of Figure A-26 suggests all places in the logical model where secondary indexing should be considered. Elements to be considered as source and target elements are listed along with a performance weight indicating the expected relative use of the index. In this

CANDIDATES FOR LOGICAL RELATIONS

Parents	Intersection
GAME-NO	GAME-NO*PLAYER-NO
PLAYER-NO	
GAME-NO	GAME-NO*TEAM
TEAM	

Figure A-25. Candidates for logical relations, iteration 3.

CANDIDATES FOR SECONDARY INDEXING

Source Element	Target Element	Weight
PLAYER-NO	GAME-NO*PLAYER-NO	400

Figure A-26. Candidates for secondary indexing, iteration 3.

case, PLAYER-NO was listed because it is the root key of at least one of the local views but is not the root of the resulting logical model.

Diagnostic Reports

As a result of the editing resolutions of the previous iterations, there are no diagnostics in iteration 3 for which action should be taken. In particular, there were no nonessential (transitive) associations in this iteration. As an exercise, the reader may wish to verify that there would have been two "apparent" nonessential associations if NO-AWARDS and NO-PENALTIES had not been renamed in iteration 2. The designer should be interested, however, in perusing a list (Figure A-27) of lone Type M associations between keys that have been augmented with Type 1 inverses.

Modified and Augmented Mappings Report

With the help of this report, the designer should determine if the assumed physical parent-child relationships, rather than logical relationships, are appropriate for the needs of the applications. Lone Type C associations between keys are included since they are treated as Type M associations.

		Forward	Inverse
	MODIFIED AND AUGMENTED MAPPINGS		
From	To	Association	Association
PLAYER-NO	POSITION-CAN-PLAY	M	1
POSITION-PLAYED	POS-CODE-1	C	1
POSITION-PLAYED	POS-CODE-2	C	1
POSITION-PLAYED	POS-CODE-3	C	1
POSITION-PLAYED	POS-CODE-4	C	1
POSITION-PLAYED	POS-CODE-5	C	1
POSITION-PLAYED	POS-CODE-6	C	1
TEAM	COACH	M	1
TEAM	PLAYER-NO	M	1
GAME-NO*PLAYER-NO	POSITION-PLAYED	M	1

Figure A-27. Modified and augmented mappings, iteration 3.

Analysis of Results

One of the unresolved editing disclosures from iteration 1 concerned the elements SCORE, HOME-SCORE, VISIT-SCORE, and WINNER. Since all four of these elements have been allocated to the same GAME-NO segment, their resolution is clearly suggested. WINNER is not needed in the data base; it can be deduced. And SCORE is not needed as a separate element. It can be the name of a group field consisting of HOME-SCORE and VISIT-SCORE.

Another unresolved editing problem involved TEAM, HOME-TEAM, VISIT-TEAM, and OPPONENT. In the GAME data base, HOME-TEAM and VISIT-TEAM appear as attributes of GAME-NO in the root segment. This seems reasonable in order to identify quickly the two teams that played in a given game. There is also a TEAM data base with TEAM as the key of its root segment. When going from GAME-NO to the TEAM data base to obtain information about a team in a particular game, OPPONENT is available as intersection data. But when going in this direction, OPPONENT is redundant because we already know, from the GAME-NO segment, which two teams played. When going in the opposite direction, from TEAM to GAME-NO, OPPONENT is still not needed as separate intersection data because a DL/I call can return to the application program the concatenation of the intersection segment and the appropriate GAME-NO segment which contains the name of the other team. Obtaining the destination parent, GAME-NO, is not to be considered an extra access because a comparison of the pertinent local views, in external views 5 and 7 with the resulting Parent-

Child graph (see Figure A-23), shows that this segment must be retrieved anyway to obtain other needed information. Therefore, the designer, with the end user's approval, decides to eliminate OPPONENT as a stored element in the GAME-NO*TEAM intersection segment. He then creates a new OPPONENTS (plural) element as a group field consisting of HOME-TEAM and VISIT-TEAM in the GAME-NO segment.

Attention is also given to the fact that TOTAL-NO-PENALTIES in the GAME-NO segment can be calculated from NO-PENALTIES in the subordinate POSITION-PLAYED segment occurrences. If the designer allows TOTAL-NO-PENALTIES to remain in the GAME-NO segment, then the only function to use it (see external view 1) can get all of its data from that one root segment. But by eliminating the TOTAL-NO-PENALTIES element and requiring the application function to calculate it from NO-PENALTIES in the POSITION-PLAYED segments, the application function would be forced to search several third-level segment occurrences to get its information. The performance weights indicate that TOTAL-NO-PENALTIES is obtained very infrequently compared to the other activity in the GAME-NO data base. Therefore, searching the third-level POSITION-PLAYED segments seems to pose a trivial performance implication. But the performance weights also show the POSITION-PLAYED segment to have heavy activity from other functions; therefore, the designer must consider the possibility of creating undesirable contention with these other functions. This will depend largely on the update requirements which we have not included. If the processing options are included with the local views as input to the automated procedures, the performance weights can be calculated to show the relative importance of the resulting paths for each type of access to be performed. Possible integrity problems of redundant storage of directly related information must also be evaluated; the Intersecting Attributes report and the Suggested Segments report can both show where redundancies occur. Finally, an independent consideration is whether or not TOTAL-NO-PENALTIES must be retained after the detail NO-PENALTIES has been deleted. As an end result, the designer decides to retain TOTAL-NO-PENALTIES in the GAME-NO segment.

With regard to the augmented Type M associations, the designer and end user determine that the parent-child relationships that will result are adequate for their current and future needs. But this is not necessarily a simple determination. Consider TEAM and PLAYER-NO, for example. The physical parent-child relationship assigns a player to one team, and player transfers from one team to another can be handled by insert and delete operations. But this means that if a player does transfer from one team to another, information relating that player to his former teams is lost. Should the end user forsee a future requirement to obtain information relating a player to any or all of his former teams, a local view of future requirements should explicitly specify a

Type M association from PLAYER-NO to TEAM. This, in combination with the Type M association already specified from TEAM to PLAYER-NO, would constitute an M:M mapping resulting in TEAM and PLAYER-NO being structured as candidates for a logical relationship rather than in a physical parent-child relationship. If the Type M association from PLAYER-NO to TEAM on the "future" local view conflicts with a different association type between these two elements on any other local view, the conflict will be revealed on an Inconsistent Associations report. To avoid adding complexity to this case study, we assume that this type of refinement is not necessary for this application or for future requirements.

ITERATION 4 LOGICAL DESIGN—REFINEMENT

Revised Data Requirements

By means of a structure specification command, it can be specified that SCORE is to be the name of a group field consisting of HOME-SCORE and VISIT-SCORE. This one command relieves the necessity of making this change in every local view containing these data element names. Similarly, OPPONENTS is made a group field consisting of HOME-TEAM and VISIT-TEAM.

With other structure specification commands, it can be specified that WINNER and OPPONENT (singular) are not to be included in the data base, and that values for both these elements are to be deduced from references to SCORE and OPPONENTS in the GAME-NO segment.

These are the only refinements to be made for iteration 4.

Reported Results

The portions of the edit, design, and diagnostic reports that are needed to understand the results of this iteration are illustrated below.

Edit Reports

Again there are no new editing problems for this iteration so the editing reports are not reproduced here.

Design Reports

All four design reports are obtained for iteration 4. These reports are described below.

Parent-Child Graph

The Parent-Child graph for iteration 4 (Figure A-28) is identical to the one obtained for iteration 3. As will be seen in the Suggested Segments report (Figure A-29), only the content of two segments has changed and the inter-segment relationships remain the same.

PARENT-CHILD GRAPH

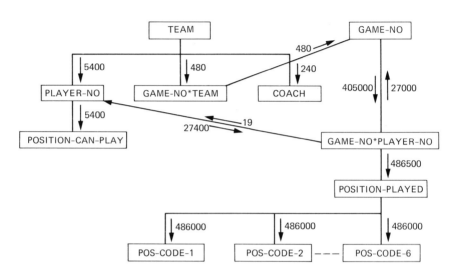

Figure A-28. Parent-Child graph, iteration 4.

Suggested Segments Report

The Suggested Segments report (Figure A-29) shows that WINNER and OPPONENT have been removed from their respective segments. These two elements are now listed in parentheses in the GAME-NO segment to indicate that, to get the information from which WINNER and OPPONENT are derived, access is to be made to this segment. In the GAME-NO segment, SCORE is now a group name for HOME-SCORE and VISIT-SCORE, and OPPONENTS is a group name for HOME-TEAM and VISIT-TEAM. The GAME-NO*TEAM segment no longer has OPPONENT as an attribute. Otherwise, the report is the same as the one in iteration 3.

Candidates for Logical Relations Report

The Candidates for Logical Relations report for iteration 4 (Figure A-30) is identical to the one for iteration 3.

SUGGESTED SEGMENTS		
Key	Attribute	Weight
COACH	COACHING-EXPERIENCE	240
	COACHING-POSITION	240
	GAMES-LOST	240
	GAMES-WON	240
	NO-COACHING-AWARDS	240
GAME-NO	SCORE	160
	HOME-SCORE	
	VISIT-SCORE	
	OPPONENTS	130
	HOME-TEAM	
	VISIT-TEAM	
	ATTENDANCE	65
	GAME-DATE	65
	LOCATION	65
	TOTAL-NO-PENALTIES	15
	(OPPONENT)	
	(WINNER)	
PLAYER-NO	PLAYER-NAME	5419
	HEIGHT	3600
	PLAYER-EXPERIENCE	3600
	WEIGHT	3600
	PRIMARY-POSITION	1819
	BIRTHDATE	1800
	TEAM-RANK	1800
	NO-INJURIES	19
POS-CODE-1 through POS-CODE-6	Attributes not listed to conserve space	486000
POSITION-CAN-PLAY	POS-RANK	5400
POSITION-PLAYED	TIME-PLAYED	486500
	NO-PENALTIES	480
TEAM	COLORS	30
	NICKNAME	30
	NO-TEAM-AWARDS	30
GAME-NO*PLAYER-NO	None	
GAME-NO*TEAM	None	

Figure A-29. Suggested segments, iteration 4.

```
┌─────────────────────────────────────────────┐
│          CANDIDATES FOR LOGICAL              │
│              RELATIONS                       │
│  ─────────────────────────────────────────   │
│     Parents            Intersection          │
│  ─────────────────────────────────────────   │
│  GAME-NO          GAME-NO*PLAYER-NO          │
│  PLAYER-NO                                   │
│  GAME-NO          GAME-NO*TEAM               │
│  TEAM                                        │
│  ─────────────────────────────────────────   │
└─────────────────────────────────────────────┘
```

Figure A-30. Candidates for logical relations, iteration 4.

Candidates for Secondary Indexing Report

The Candidates for Secondary Indexing report for iteration 4 (Figure A-31) is identical to its counterpart in iteration 3.

```
┌──────────────────────────────────────────────────────────────┐
│          CANDIDATES FOR SECONDARY INDEXING                   │
│  ──────────────────────────────────────────────────────────   │
│  Source Element       Target Element          Weight         │
│  ──────────────────────────────────────────────────────────   │
│   PLAYER-NO        GAME-NO*PLAYER-NO            400           │
│  ──────────────────────────────────────────────────────────   │
└──────────────────────────────────────────────────────────────┘
```

Figure A-31. Candidates for secondary indexing, iteration 4.

Diagnostic Reports

As a result of the refinements made for this iteration, one nonessential (transitive) association now exists. It is reported in Figure A-32.

```
┌──────────────────────────────────────────────────────────────┐
│       NONESSENTIAL (TRANSITIVE) ASSOCIATIONS                 │
│  ──────────────────────────────────────────────────────────   │
│  Association Removed                            Type         │
│      GAME-NO*TEAM  →  OPPONENTS                  1           │
│  Alternate Path:     GAME-NO*TEAM  →                        │
│                      GAME-NO  →                             │
│                      OPPONENTS                              │
│  ──────────────────────────────────────────────────────────   │
└──────────────────────────────────────────────────────────────┘
```

Figure A-32. Nonessential (transitive) associations, iteration 4.

Nonessential (Transitive) Associations Report

Analysis of Results

The Nonessential Association report shows that the association from GAME-NO*TEAM to OPPONENTS was removed from the logical model and that access must be made to GAME-NO to determine OPPONENTS. This is acceptable because OPPONENTS is an attribute in the GAME-NO segment.

The designer is satisfied with the logical design as it now stands and the end user is satisfied that it fully supports the nine initial functional requirements.

ITERATION 5 LOGICAL DESIGN—ADDING ADDITIONAL REQUIREMENTS

Additional Data Requirements

Now that a logical design has been determined for the nine functions of the initial applications, the requirements of another application of lesser importance will be added to the design study. The purpose of further iterations of the design procedures is to determine if the existing logical design will accomodate the new application functions, and if not, to show where trade-offs should be considered and to give supporting diagnostics to help the designer evaluate those trade-offs. One new function, described by external view 10 (see Figure A-33), will be added in this iteration.

Reported Results

The reports of interest from iteration 5 are discussed below.

Edit Reports

Again the edit reports have nothing new to reveal, but this will not necessarily be the case in a real design study. These reports should always be reviewed.

Design Reports

The four design reports for iteration 5 follow.

Parent-Child Graph

The Parent-Child graph that results from iteration 5 (Figure A-34) is identical to the one in iteration 4.

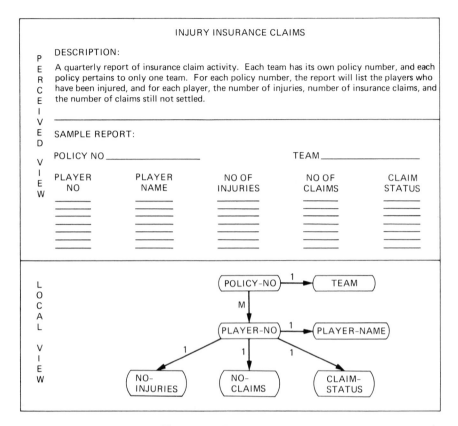

Figure A-33. External view 10.

Suggested Segments Report

The Suggested Segments report for iteration 5 (Figure A-35) is the same as that for iteration 4 except that POLICY-NO is included as an attribute in the TEAM segment, and CLAIM-STATUS and NO-CLAIMS are added to the PLAYER-NO segment.

Candidates for Logical Relations Report

The Candidates for Logical Relations report for iteration 5 (Figure A-36) is identical to the ones for iterations 3 and 4.

Candidates for Secondary Indexing Report

The candidates for secondary indexing have changed. The report (Figure A-37) now lists POLICY-NO as a possible source element with PLAYER-NO and TEAM as its targets.

PARENT-CHILD GRAPH

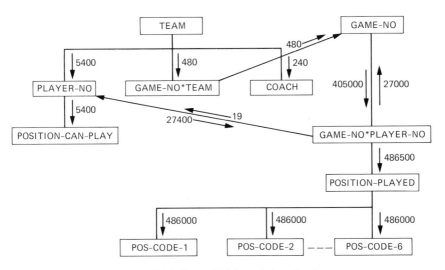

Figure A-34. Parent-Child graph, iteration 5.

Diagnostic Reports

The addition of the new requirement has created some design alternatives that must be recognized and evaluated by the designer. The following diagnostic reports will help the designer recognize these situations and will provide information to help in the evaluation.

Key-Only Segment Resolutions Report

This report shows that a segment having only a key element and no attributes was derived by the automated procedures.

Modified and Augmented Mappings Report

This report (Figure A-39) is the same as that obtained in iteration 3 except that an entry from POLICY-NO to PLAYER-NO has been added.

Analysis of Results

A glance at the Parent-Child graph (see Figure A-32) and the Suggested Segments report (see Figure A-33) indicate that a very innocuous change has been made by adding the new requirement. POLICY-NO is added as an attribute to the TEAM segment and no change is made to the hierarchical structure.

SUGGESTED SEGMENTS

Key	Attribute	Weight
COACH	COACHING-EXPERIENCE	240
	COACHING-POSITION	240
	GAMES-LOST	240
	GAMES-WON	240
	NO-COACHING-AWARDS	240
GAME-NO	SCORE	160
	HOME-SCORE	
	VISIT-SCORE	
	OPPONENTS	130
	HOME-TEAM	
	VISIT-TEAM	
	ATTENDANCE	65
	GAME-DATE	65
	LOCATION	65
	TOTAL-NO-PENALTIES	15
	(OPPONENT)	
	(WINNER)	
PLAYER-NO	PLAYER-NAME	5419
	HEIGHT	3600
	PLAYER-EXPERIENCE	3600
	WEIGHT	3600
	PRIMARY-POSITION	1819
	BIRTHDATE	1800
	TEAM-RANK	1800
	CLAIM-STATUS	58
	NO-CLAIMS	58
	NO-INJURIES	58
POS-CODE-1 through POS-CODE-6	Attributes not listed to conserve space	486000
POSITION-CAN-PLAY	POS-RANK	5400
POSITION-PLAYED	TIME-PLAYED	486500
	NO-PENALTIES	480
TEAM	COLORS	30
	NICKNAME	30
	NO-TEAM-AWARDS	30
	POLICY-NO	0
GAME-NO*PLAYER-NO	None	
GAME-NO*TEAM	None	

Figure A-35. Suggested segments, iteration 5.

CANDIDATES FOR LOGICAL RELATIONS	
Parents	Intersection
GAME-NO	GAME-NO*PLAYER-NO
PLAYER-NO	
GAME-NO	GAME-NO*TEAM
TEAM	

Figure A-36. Candidates for logical relations, iteration 5.

CANDIDATES FOR SECONDARY INDEXING		
Source Element	Target Element	Weight
PLAYER-NO	GAME-NO*PLAYER-NO	400
POLICY-NO	PLAYER-NO	58
POLICY-NO	TEAM	2

Figure A-37. Candidates for secondary indexing, iteration 5.

KEY-ONLY SEGMENT RESOLUTIONS	
Key of Removed Segment	Key of Receiving Segment
POLICY-NO	TEAM

Figure A-38. Key-Only segment resolutions, iteration 5.

Although this is the suggested result, the situation is somewhat more complex than this and, as shown by the diagnostic reports, requires careful evaluation.

Looking first at the Key-Only Segment Resolutions report (see Figure A-38), we see that POLICY-NO was the key of a segment that had no attributes and that the automated procedures removed this key-only segment by placing POLICY-NO as an attribute into the TEAM segment, according to the rules developed in Chapter 9. This result is shown in the Suggested Segments report (see Figure A-33). Had this action not been taken by the automated procedures, the POLICY-NO segment would have been structured between the TEAM and the PLAYER-NO segments as shown in the Alternate Parent-Child Graph of Figure A-40. (By analyzing the Type 1 associa-

MODIFIED AND AUGMENTED MAPPINGS

From	To	Forward Association	Inverse Association
PLAYER-NO	POSITION-CAN-PLAY	M	1
POLICY-NO	PLAYER-NO	M	1
POSITION-PLAYED	POS-CODE-1	C	1
POSITION-PLAYED	POS-CODE-2	C	1
POSITION-PLAYED	POS-CODE-3	C	1
POSITION-PLAYED	POS-CODE-4	C	1
POSITION-PLAYED	POS-CODE-5	C	1
POSITION-PLAYED	POS-CODE-6	C	1
TEAM	COACH	M	1
TEAM	PLAYER-NO	M	1
GAME-NO*PLAYER-NO	POSITION-PLAYED	M	1

Figure A-39. Modified and augmented mappings, iteration 5.

tions into and out of POLICY-NO in external view 10, the reader can verify this result.) A Nonessential (Transitive) Association report of Figure A-41 would have been obtained to alert the designer that the direct association from TEAM to PLAYER-NO has been replaced with a path from TEAM to POL-ICY-NO to PLAYER-NO. By comparing the performance weight of 5400 for this direct path that has been removed to 0 and to 58 respectively for the path from TEAM to POLICY-NO and from POLICY-NO to PLAYER-NO, the designer would see a severe performance implication in placing POLICY-NO between TEAM and PLAYER-NO and he would want to consider some alternatives. But in this case, the automated procedures recognize that POLICY-NO has no defined attributes and therefore eliminate it as a segment.

In many cases, placing a lone key such as POLICY-NO into the segment of its parent will be a reasonable and desirable solution. But in this particular case, the designer should recognize that POLICY-NO and PLAYER-NO pertain to two different entities and he should question the wisdom of putting them into the same segment. Having them together may be more efficient for some kinds of processing, but it could lead to contention between programs and also make the data base less adaptable to program changes or to additional requirements. Attempts to impose intelligent constraints on the natural structuring process are always reported to the designer for his evaluation.

Finally, in either case, the designer should examine the Modified and Augmented Mappings report (See Figure A-39) to see if the Type 1 inverse associations, which were automatically supplied for lone Type M associations between keys and induce parent-child relationships, are adequate for current

PARENT-CHILD GRAPH

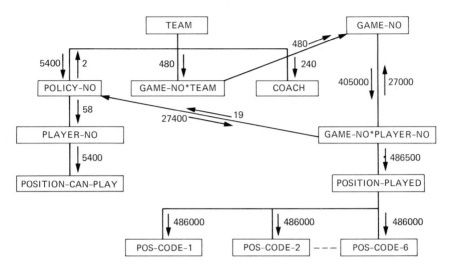

Figure A-40. Parent-Child graph, iteration 5 (alternate).

NONESSENTIAL (TRANSITIVE) ASSOCIATIONS

Association Removed	Type	Weight
PLAYER-NO → TEAM	1	0
Alternate Path: PLAYER-NO → POLICY-NO → TEAM		
Association Removed	Type	Weight
TEAM → PLAYER-NO	M	5400
Alternate Path: TEAM → POLICY-NO → PLAYER-NO		
Association Removed	Type	Weight
GAME-NO*TEAM → OPPONENTS	1	480
Alternate Path: GAME-NO*TEAM → GAME-NO → OPPONENTS		

Figure A-41. Nonessential (transitive) associations, iteration 5 (alternate).

and future needs. The designer and end user decide to accept POLICY-NO as an attribute of the TEAM-NO segment pending further study.

The designer and the end user are now ready to study the effect that future data requirements may have on the logical design thus far derived. The future requirements that they envision are described in iteration 6.

ITERATION 6 LOGICAL DESIGN—ADDING FUTURE REQUIREMENTS

The tendency to try to create a key-only segment of POLICY-NO in the previous iteration causes the designer to question the completeness of this new data requirement as submitted. Even though the requirements as submitted support the needs of the current application functions, perhaps more should be specified to accomodate possible future requirements. Usually, there will be attributes (description, expiration date, etc.) about an insurance policy. (Here we are really describing current intrinsic requirements and future functional requirements.) But the data requirements were based on the concept of one policy per team and vice versa.

Looking into the future, the designer and end user agree that the time may come when a team is covered by more than one type of policy, and each type will have its own description, expiration date, and so forth. As a consequence, each player may be covered by more than one type of policy. The designer and end user decide to enter these revised relationships into the design study to see what the implications will be. They would like to accomodate these future requirements if they can do so without adversely impacting the quality of support for their current requirements. And if conflicts are created by the future requirements, the designer wants to know where they will be. Design iterations of this type are made feasible when using automated procedures to process machine-readable data requirements.

Future Data Requirements

The future requirements, as described above, result in four new associations shown in Figure A-42.

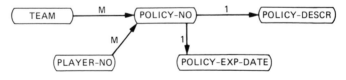

Figure A-42. Future requirements.

These futuristic associations can be entered into the design study as a separate local view and processed in a separate iteration in order to observe any conflicts that may arise. This is the approach to be taken. Howeve, for illustrative purposes, the four new associations in Figure A-43 are superimposed onto the local view of external view 10.

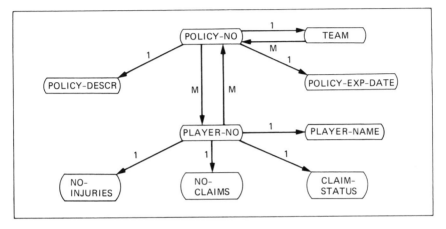

Figure A-43. Augmented local view 10.

Reported Results

The results of interation 6, in abbreviated form, are reported below.

Edit Reports

No additional editing problems are reported in this iteration.

Design Reports

For brevity, we will show only the resulting Parent-Child graph in Figure A-44. It shows that POLICY-NO again defines a segment. But this time, POLICY-NO is a fourth child of TEAM, and PLAYER-NO is still a direct child of TEAM. The Suggested Segments report, if illustrated, would show a POLICY-NO segment with POLICY-DESCR and POLICY-EXPR-DATE as attributes. Otherwise, it is the same as that illustrated in Figure A-33. The Candidates for Logical Relations report will show an additional entry because of the M:M mapping that now exists between PLAYER-NO and POLICY-NO. Otherwise, it will be the same as that shown in Figure A-35. The Can-

PARENT-CHILD GRAPH

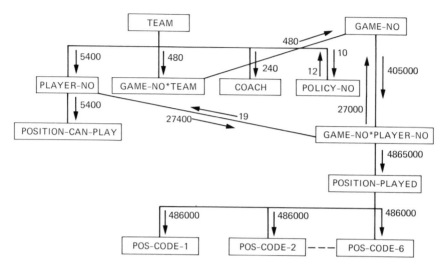

Figure A-44. Parent-Child graph, iteration 6.

didates for Secondary Indexing reports will be the same as those shown in Figure A-36.

Diagnostic Reports

Although the edit and diagnostic reports should always be printed and examined by the designer, they are not illustrated here because they are similar in content to what we have already seen and will not show anything requiring further action.

Analysis of Results

The designer and end user are happy with these results. Accesses from TEAM to PLAYER-NO are not degraded by having a new segment between them and policy information, which describes a different kind of entity than player, is no longer mixed with player information in the TEAM segment. Even if there is only one POLICY-NO segment occurrence for each TEAM segment occurrence, this can be a desirable result. In fact, if this result had not been obtained automatically, the designer probably would have specified a refinement of this type and then processed it in an additional iteration as a quality control check to see if any significant new diagnostics occurred.

The designer chooses to implement both candidates for logical relationships, as indicated in the Parent-Child graph of Figure A-44, in order to have the required access paths. But he chooses not to implement the suggested logical relation between POLICY-NO and PLAYER-NO because, since they are both children of TEAM, the necessary paths for the functional requirements are present. The designer also chooses not to implement the suggested secondary index from PLAYER-NO to GAME-NO*PLAYER-NO. One of the logical relations provides this path, and the performance weights show it to be traversed relatively infrequently. The logical design is now considered to be completed.

ALTERNATIVE LOGICAL MODELS FOR PHYSICAL DESIGN EVALUATION

Even though a logical model has been obtained, the designers want to know how well it will perform, and if there are variations of the logical model that might perform better in a production environment where contention and the properties of access method, pointer options, etc., can be considered. Some of the more obvious possible variations are:

- A separate policy data base with POLICY-NO logically related to TEAM
- A secondary index from PLAYER-NO to GAME-NO
- Removal of TOTAL-NO-PENALTIES from the GAME-NO segment and calculating its value by references to NO-PENALTIES in the subordinate POSITION-PLAYED segment occurrences

After selecting the various physical storage options, the design variations noted above can be evaluated as part of the physical design process. Our purpose here was merely to illustrate the use of computer-assisted techniques for the conceptual and logical design phases, and thus this case study will not extend into the physical design realm. Application of automated techniques for physical design evaluation has been reviewed in Part III of this book and is fairly well known within the community of DL/I data base designers.

SUMMARY

In this case study we have tried to illustrate the concepts of computer-assisted data base design. In so doing, we have emphasized the iterative nature of the process and the human control that must be exercised. One of the major contributions of these procedures is in helping to identify the editing problems that exist when the data requirements specified for different applications are brought together. Other major contributions are the wealth of diagnostic information obtained in the logical design phase and the evaluations performed in physical design.

We have indicated the kinds of help that are provided by the editing procedures, but we have touched only lightly on the help offered by the diagnostic reports in the logical design phase. In a design study of real-life proportions, the diagnostic reports will provide much valuable understanding of alternatives, inconsistencies, and problems detected in the structuring process. By providing such editing and diagnostic information early in the design process, and by enabling repetitive iterations from machine readable data requirements, these automated procedures should materially assist the designer in obtaining an efficient design for current requirements and future requirements as he is able to forsee them and in reducing the time required to complete the design study.

Appendix B
Professional Reader's Review and Analysis

CHAPTER 2

1. What are the fundamental differences between identifiers, keys, and attributes?
2. How do associations differ from mappings? Why is the distinction important when specifying data requirements?

CHAPTER 3

1. What are the three views of data, and what are the major characteristics of each?
2. Which of these views correspond closely with the contents of a dictionary system? In what ways?
3. How may the automated design procedures be used so as to help in identifying and evaluating trade-offs between conflicting requirements of local views or between current and future requirements?

CHAPTER 4

1. What things can a composite model reveal that will not be obvious in the individual local views?
2. What are the relative merits of having the end users or the data base designers specify the initial data requirements?
3. Comment on the statement: The real design work is in determining the proper association types between related elements.

CHAPTER 5

1. What are the relative roles of the human designer and the automated procedures in the editing process? Is there such a thing (in the current state of the art) as automatically editing the data requirements? What is the real contribution of the automated procedures to editing?
2. How might more consideration of the semantic meaning of the data be built into the editing process?

CHAPTER 6

1. In the following four composite models, which elements will be keys, which will be attributes, and what segments will be produced? Assume that the automated procedures will supply Type 1 inverses for lone Type M associations between keys.

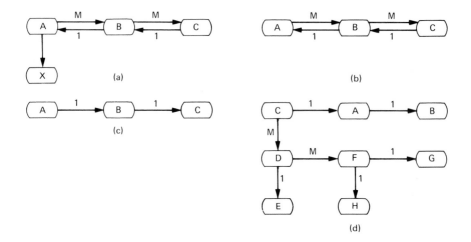

(a)   (b)   (c)   (d)

2. What structures will each of the following composite models produce? Assume that the automated procedures will supply Type 1 inverses for lone Type M associations between keys.

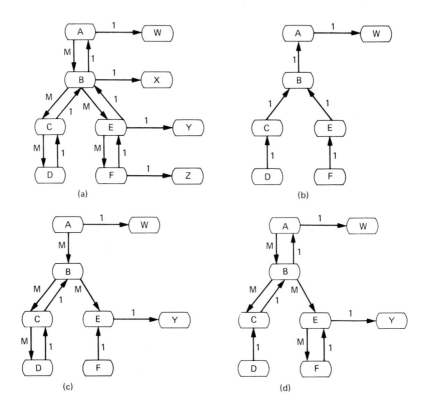

(a)   (b)   (c)   (d)

CHAPTER 7

1. In each of the following bubble charts, a logical relationship will be suggested by the automated procedures. Which could be supported by a undirectional logical relationship, and which will require a bidirectional logical relationship? Assume in all cases that A, B, and C are keys, although their attributes are not illustrated.

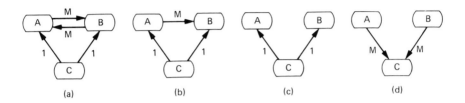

2. If the following local views are combined into a composite model, two candidates for logical relationships will be derived. What will be the keys of the intersection segments?

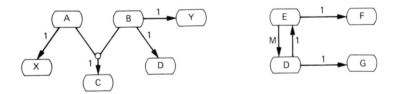

3. From the following bubble chart, some secondary indexes will be suggested. Which will be the source and target elements in each?

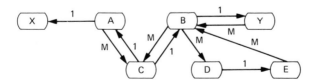

4. Each of the following composite models contains one or more compound keys. What structures will be produced? Assume in all cases that A, B, and C are keys, although their attributes are not illustrated.

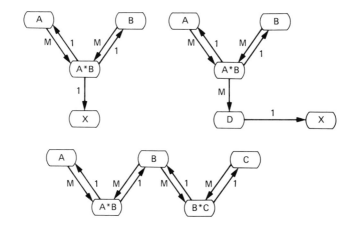

CHAPTER 8

1. In which of the folowing cases will the Type C association be treated as Type 1 or as Type M?

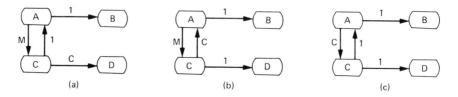

2. In each of the following composite models, determine the contents of the root segments that will be derived. Assume that the automated procedures will supply Type 1 inverses for lone Type M associations between keys.

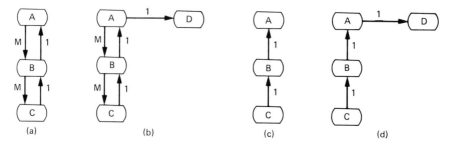

3. If the following local views are combined into a composite model, a candidate for a logical relationship will be derived. What will be the keys of the intersection and of the physical and logical parents as implemented?

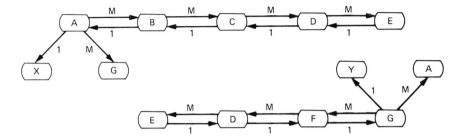

4. If the following local views are combined into a composite model, a loop will exist. Determine the elements that will constitute the loop.

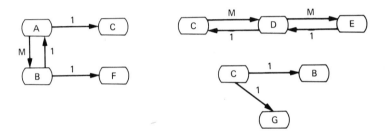

CHAPTER 9

1. In the following composite models, which elements will initially be classed as keys, and which key-only segments will be eliminated? Draw the structures that will be produced by the automated procedures.

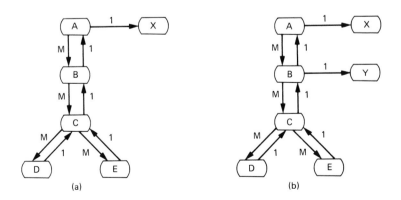

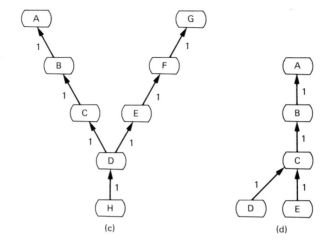

(c)   (d)

2. In the following composite models, compound keys are specified. Which of the compound keys will be unnecessary in the resulting structures? Draw the resulting structures. Assume in all cases that A, B, and C are keys, although their attributes may not be illustrated.

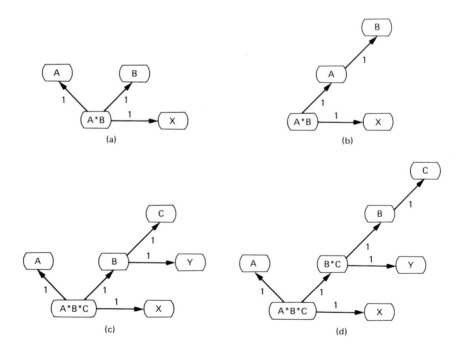

3. How will each of the following identities be resolved? Draw the structures that the automated procedures will produce.

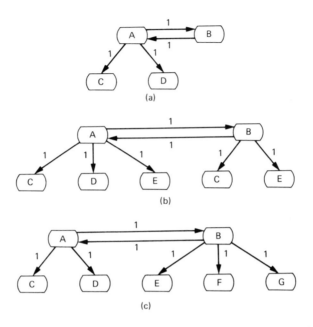

(a)

(b)

(c)

CHAPTER 10

1. In the following local views, what will be the resultant frequency of use of each path? The number (n) between element pairs means that for each access to an occurrence of the "from" element there will be an average of n accesses to occurrences of the "to" element.

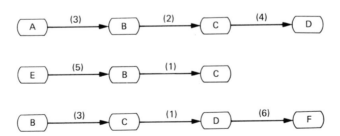

2. If the local views in (1) are combined into a composite model, what will be the resultant frequency of use of each of the resulting paths? Assume they are already on a common time and setup basis.

CHAPTER 12

1. Name some reasons for refining the canonical logical model produced by the automated procedures.
2. One of the main roles of the automated procedures in deriving the canonical logical model is to organize the data requirements. When the designer dictates refinements to the logical model, one of the main roles of the procedures becomes quality control. What are the significant features and values of these roles?
3. With machine–readable data requirements, what are some ways in which the iterative automated procedures can be used to reveal design trade–offs and to consider future requirements?

CHAPTER 13

1. Are a minimum covering of third-normal form relations necessarily the ones to be implemented? Give reasons for your answer. What are the advantages of obtaining minimum coverings as part of the design process?
2. Fourth-normal form relations imply an understanding of the semantics of the data requirements beyond that which is required for obtaining third-normal form. Describe this additional understanding. What ways can you suggest for incorporating it into the automated procedures?

CHAPTER 14

1. CODASYL data bases are said to be network structures; DL/I data bases, as implemented with logical relations, can be described as "restricted" networks. What are the major differences between these two network concepts?
2. What are labeled associations? How are they implemented in CODASYL? In what ways can they be implemented in DL/I?

CHAPTER 15

1. In a newly loaded or reorganized data base, HISAM may perform better than HIDAM for the same data base, but after a period of updating, HISAM usually becomes slower than HIDAM. What are some of the reasons for this degradation?
2. Which major DL/I access method does not permit updating?
3. An advantage of HDAM over HIDAM is that file accesses are not required to search an index. What techniques can be employed with HIDAM to minimize this difference?

CHAPTER 16

1. Which DL/I storage organization is usually used for volatile data bases? Why?
2. What are some reasons for dividing a data base into secondary data set groups?

3. What are the performance implications when several keys are randomized to the same storage address?

CHAPTER 17

1. Why is the average record sometimes not a good representation of the data base records for space (and time) calculations?
2. What choices of free space are permitted by each of the major DL/I access methods?
3. In what ways is the expected amount of waste space influenced by shorter blocks and by longer blocks?

CHAPTER 18

1. Increasing the block size decreases the likelihood of I/O when accessing a segment. What are some of the reasons for restricting the sizes of blocks?
2. What must be known about the application program's use of the data base in order to draw useful conclusions from the I/O probabilities?

CHAPTER 19

1. In what ways is modeling advantageous over the analytic space and time calculations? In what ways are the analytic calculations advantageous?
2. In what ways can the development of application models assist in the development of production application programs?

References

The literature abounds with articles on various facets of conceptual design. Modeling, languages for describing the real world, data description languages, processing languages, equivalences and differences between data structures, and other study areas are treated. But very little has been published about concepts, techniques, and procedures for doing the design and for evaluating it once it is done. In these areas, the following references are recommended as supplementary reading.

A. GENERAL

1. Engles, R. W., A Tutorial on Data Base Organization, *IBM Technical Report* TR00.2004 (1970).
2. Yormark B., The ANSI/SPARC/SGDBMS Architecture, *The ANSI/SPARC DBMS Model,* North-Holland (1977).
3. *CODASYL Data Base Task Group Report,* ACM (1971).
4. Date, C. J., *An Introduction to Database Systems,* Addison-Wesley (1977).
5. Martin, J., *Principles of Data-Base Management,* Prentice-Hall (1976).
6. McElreath, T. J., *IMS Design & Implementation Techniques,* Q.E.D., Information Sciences, Inc. (1979).
7. Fagin, R., Multivalued Dependencies and a New Normal Form for Relational Databases, *IBM Research Report* RJ 1812 (July, 1976).
8. Codd, E. F., A Relational Model for Data for Large Shared Data Banks, *Communications of the ACM* **13**:377–387 (June, 1970).
9. Olle, T. W., *The CODASYL Approach to Data Base Management,* John Wiley & Sons (1978).
10. Cardenas, A. F., *Data Base Management Systems,* Allyn and Bacon (1979).
11. Smith, J. M. and Smith, D. C. P., Principles of Database Design, *Proceedings of the New York University Symposium on Data-Base Design,* New York University (1978).
12. Palmer, I., Practicalities in Applying a Formal Methodology to Data Analysis, *Proceedings of the New York University Symposium on Data-Base Design,* New York University (1978).

B. CONCEPTUAL DESIGN PROCEDURES

1. Brown, A. P. G., Modeling a Real World Process, *Data Base Description,* North-Holland (1975).
2. Adida, M., Delobel, C., and Leonard, M., A Unified Approach for Modeling Data in Logical Data Base Design, *Modeling in Data Base Management Systems,* North-Holland (1976).
3. Gane, C. and Sarson, T., *Structured Systems Analysis: Tools and Techniques,* Improved Systems Technologies Inc. (1977).
4. Martin, J., *Computer Data-Base Organization,* Prentice-Hall (1977).

C. LOGICAL DESIGN PROCEDURES

1. Raver, N. and Hubbard, G. U., Automated Logical File Design, *IBM Systems Journal* 16, 3:287–312 (1977).
2. Hubbard, G. U., A Technique for Automated Logical Data Base Design, *Proceedings of the New York University Symposium on Data Base Design,* New York University (1978).
3. IBM Corporation, *Data Base Design Aid (Version 2): Designer's Guide,* Publication No. GH20-1627 (1977).
4. Kroenke, D., *Database Processing,* Science Research Associates (1977).
5. Wang, C. P. and Wedekind, H. H., Segment Synthesis in Logical Data Base Design, *IBM Journal of Research and Development* 19, 1:71–77 (Jan. 1975).
6. Wang, C. P., Parameterization of Information System Applications, *IBM Research Report* RJ-1199 (April 1973).
7. Bernstein, P. A., Synthesizing Third Normal Form Relations from Functional Dependencies, *ACM Transactions on Database Systems* I, 4:277–298 (Dec. 1976).
8. Selinger, P. G., Astrahan, M. M., Chamberlin, D. D., Lorie, R. A., and Price, T. G., Access Path Selection in a Relational Database Management System, *Proceedings of the ACM-SIGMOD International Conference on Management of Data* 23–34 (1979).
9. Sheppard, D., Principles of Data Structure Design, *Auerbach Database Management Series,* Portfolio Number 23-01-04 (1977).
10. Dyba, E., Principles of Data Element Identification, *Auerbach Database Management Series,* Portfolio Number 23-01-03 (1977).
11. Mitoma, M. F., "Optimal Database Schema Design," Ph.D. Dissertation, University of Michigan (1975).
12. Chen, P. S., The Entity-Relationship Model: Toward a Unified View of Data, *ACM Transactions on Database Systems* (March, 1976).
13. Smith, J. M. and Smith, D. C. P., Database Abstractions: Aggregation and Generalization, *ACM Transactions on Database Systems* (June, 1977).
14. Beeri, C. and Bernstein, P. A., Computational Problems Related to the Design of Normal Form Relational Schemes, *ACM Transactions on Database Systems* (July 1976).
15. Bernstein, P. A., Comment on Segment Synthesis in Logical Data Base Design, *IBM Journal of Research and Development* 20, 4:112 (July 1976).

D. PHYSICAL DESIGN PROCEDURES

1. Kapp, D. and Leben, J., *IMS Programming Techniques: A Guide to Using DL/I,* Van Nostrand Reinhold (1978).
2. Dechow, E. and Lundberg, D., The IMS Data Base Application Design Review, *IBM Report* G320-6009 (1977).
3. *The Data Base Design Guide,* GUIDE International Corporation (1974).
4. IBM Corporation, *DBPROTOTYPE/II Program Description/Operations Manual,* Publication No. SH20-1953 (1978).
5. IBM Corporation, *IMS/VS Version 1 System/Application Design Guide,* Publication No. SH20-9025 (1976).
6. IBM Corporation, *IMS/VS Utilities Reference Manual,* Publication No. SH20-9029 (1976).

Index

Access Methods (IMS), 51, 137–149, 160
  comparison, 146–148
  HDAM, 137–139, 143–156, 176–177,
    183, 184
  HIDAM, 138–139, 143–148, 150–156,
    176
  HISAM, 137–144, 146–148, 150–155,
    176, 184
  HSAM, 137–141, 146, 150–153
  selection of, 150–152
Access Methods (OS)
  BSAM, 138–139
  ISAM, 22, 138–139, 141–142, 147
  OSAM, 138–139, 141–142, 147
  QSAM, 138–139
  VSAM (ESDS), 138–139, 141, 147, 153
  VSAM (KSDS), 138–139, 141, 147, 153
Accessing Frequencies, 32. *See also* Performance weights
Aggregation principle, 100
Application program modeling, 187–195.
    See also Modeling, application
    program
Association paths, 60, 62, 65, 68, 77–78,
    80–81, 84–86, 101, 122
Associations, 12–13, 65, 101, 207, 230–232
  inconsistent, 45–46, 133, 217–218, 228,
    230, 237–239
  inverse, 55–56, 64, 77, 96, 117, 119,
    130–131, 260
  labeled, 129–130
  lone Type M, 55–57, 64–65, 116–117,
    119, 130–131, 133
  lone Type 1, 75–79
  nonessential, 57, 59, 95, 101, 104–105
  transitive, 57–59, 95, 117–120, 124, 128
  vs. mappings, 12–13
Association types, 30–32, 207, 230–232
  complex (Type M), 13–14, 27–32, 35,
    45, 55–57, 63–68, 69, 94–96, 106,
    116, 121, 131, 133, 210, 239–240,
    250–251
  conditional (Type C), 14, 69, 96

simple (Type 1), 13, 27–32, 35, 45, 53–
    57, 60–62, 64, 66, 68–69, 96, 106,
    116–117, 119, 121, 130–133, 210,
    239, 260
Atomic element, 59, 114, 116, 134
Attributes, 11, 30, 53, 55–57, 60, 62, 66–
    68, 69–71, 81–84, 100, 105–107,
    122–124, 131, 216
  dummy, 70
  floating, 70–71, 105–106
  intersecting, 46–47, 96, 232–233, 237,
    239
  mutually independent, 116–118
  repeating, 70–71, 105–106
ANSI-SPARC, 9

Bernstein, P. A., 121, 123–124
Binary relations, 27, 41–42, 54, 118, 122.
    See also Bubble charts
  deriving, 29–30
Block, 141–146. *See also* End of block
    waste
  optimum size, 152–153
  overflow, 144–146
  root addressable, 144–146, 149, 160,
    188
  size, 3, 51, 143, 145, 150, 160, 188.
    See also Effective block size
Blocking factor, 143
Bottom-up structuring, 54, 55–62, 131
Bubble charts, 27, 33, 41, 85–87. *See also*
    Binary relations
Buffer pool statistics, 187, 191, 195
Byte limit, 144, 146

Calculations
  space, 160–169
  time, 160, 170–186
Call patterns. *See* DL/I call patterns
Candidate
  key, 11
  logical designs, 135